APPLIED HIGH-RESOLUTION
GEOPHYSICAL METHODS

Offshore Geoengineering Hazards

APPLIED HIGH-RESOLUTION GEOPHYSICAL METHODS

Offshore Geoengineering Hazards

Peter K. Trabant, Ph.D.

 Springer-Science+Business Media, B.V.

Library of Congress Cataloging in Publication Data

Trabant, Peter K., 1942–
    Applied high-resolution geophysical methods.

    Includes bibliographical references and index.
    1. Echo sounding. 2. Offshore structures—Safety measures.
3. Drilling platforms—Safety measures. 4. Engineering
geology—Safety measures. I. Title. GC78.E3T7   1984
627'.9   83-26662

ISBN 978-94-009-6495-2   ISBN 978-94-009-6493-8 (eBook)
DOI 10.1007/978-94-009-6493-8

Frontispiece: "Artist's Conception of the Mid-Pacific
Mountains." From Hamilton, E.L., 1956, Sunken islands of
the mid-Pacific mountains: Geol. Soc. Am., Memoir no. 64,
plate 1, 97 p. Reprinted by permission. Oil painting by
Chesley Bonestell.

*Dedicated to the memories of Dave Harvey, Joe Gray,
and Ken Cram*

CONTENTS

ACKNOWLEDGMENTS

The author wishes to express his deepest thanks to the following persons, and companies, for their support and assistance during the compilation of this text: Dr. James L. Harding, Ph.D., Director, Marine Extension Service, University of Georgia, and Mr. Lloyd Stahl, consultant geophysicist, COMAP Geosurveys, Inc., for their review of major portions of the text.

Mr. Martin Klein of Klein Associates, Inc.; Bill Jones, Ph.D., of B. R. Jones and Associates; Jean Audibert, D.Sc., of the Earth Technology Corp.; Mr. Colin G. Weeks, F.R.I.C.S., of Wimpol Ltd.; Mr. Kelly Robertson, and Clyde Lee of Sytech Corp., for review of individual chapters.

ARCO Oil and Gas Company; McClelland Engineers, Inc.; Mr. Jack Hudson of Cultural Resource Services, Inc.; Mr. Mark Lawrence of Fairfield Industries, Inc.; Marine Technical Services, Inc.; Exxon Production Research Company; Oceanonics, Inc.; COMAP Geosurveys, Inc.; C. A. Richards and Associates, Inc.; Hunt Oil Company; and Mesa Petroleum Company of Houston, Texas. The Raytheon Company of Portsmouth, Rhode Island; Odom Offshore Surveys, Inc.; and Coastal Environments, Inc. of Baton Rouge, Louisiana. Klein and Associates of Salem, New Hampshire; Nekton, Inc. of San Diego, California; Dames and Moore Consultants of Cranford, New Jersey; and BEICIP and CGG of France.

My most grateful appreciation is extended to both my wife Virginia (Sissy) and "Ozzie," my portable Osborne 1 microprocessor (#588), who unrelentingly supported my efforts throughout this production.

ACKNOWLEDGMENTS

One
INTRODUCTION

The discipline encompassing the use of high-resolution geophysics for obtaining geoengineering survey data has evolved rapidly over the past decades to become an interdisciplinary subject encompassing the fields of Geophysics, engineering, geology, marine geology, oceanography, and civil engineering. While high-resolution geophysical surveys are routinely performed offshore today, this has been so only since the late 1960s.

High-resolution geophysical methods are employed in the offshore environment to obtain a comprehensive picture of the sea-floor morphology and underlying shallow stratigraphy. The purpose of the survey methods is to assist in the design and installation of bottom-supported structures such as drilling and production platforms and pipelines.

Drilling structures and pipelines of steel and/or concrete have become behemoths with respect to their size and the complexity of their design in order to withstand, for periods of up to twenty-five years, an extremely harsh environment, including storm waves, strong currents, unstable sea-floor conditions, and great water depths. It is therefore of paramount importance that the geometry and physical properties of the sea floor be well understood in order to provide an adequate foundation for the design lives of such structures.

On land, engineering foundation data usually may be obtained by visual field inspection and shallow borehole information, but offshore the presence of the water column places certain constraints on geoengineering investigations.

High-resolution geophysical methods employed in the acquisition of geoengineering data offshore are defined as the use of seismic sources and receivers that operate at acoustic frequencies greater than 100 Hz.

A comprehensive marine geoengineering survey requires the deployment of a number of acoustic tools from a precisely positioned survey vessel. Equipment employed includes echo sounders, side-scanning sonars, subbottom profilers, sparkers, or other deeper penetrating acoustic sources, as well as total field intensity or gradient magnetometers and appropriate positioning equipment. The interpretation of such survey data permit the assessment of geological and other features both on and beneath the sea floor, some of which may constitute either hazards or constraints to the placement of bottom-supported or penetrating engineering structures.

Frequently referred to as hazards, such features may include faults, shallow accumulations of overpressured gas, the presence of man-made debris such as lost drilling equipment or sunken vessels, or lateral variations of the physical properties of the sea floor such as occur in association with infilled relict river channels. Lack of attention to such

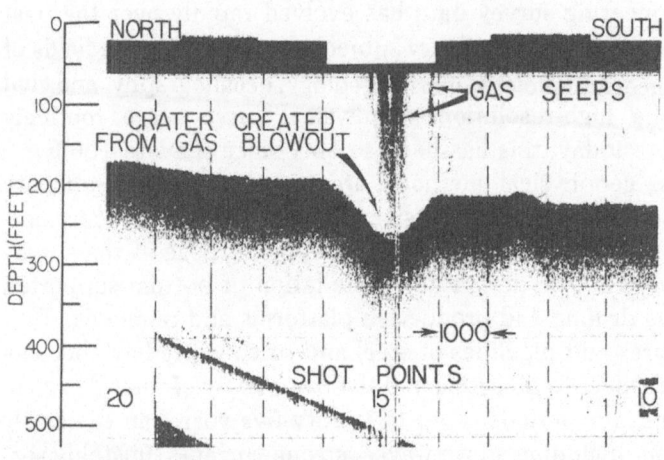

FIGURE 1.1 3.5 kHz subbottom profiler record over crater produced by shallow gas blowout. (From Antoine, J., and P. Trabant, 1976, Geological features of shallow gas, *in* Exploration and engineering high resolution geophysics: Continuing Education Symposium by the Geophysical Society of Houston. By permission of the authors.)

features may lead to the loss of a drilling platform by a *blowout* and the escape of large volumes of hydrocarbons into the marine environment. The high-resolution seismic profile of figure 1.1 illustrates the result of a blowout where a drilling platform was lost.

HISTORICAL BACKGROUND
The use of the lead line for obtaining water depths and indications of the nature of the sea floor has been prevalent since prehistoric times. Herodotus, the Greek father of history, wrote of soundings taken on the Nile, while an Egyptian artist of the Twelfth Dynasty (1991 to 1786 B.C.) depicted the earliest recording of the lead sounding technique (1).

It was during the nineteenth century that the lead line made of hemp was replaced by steel wire spooled on a mechanical winch, to provide deeper and faster soundings. The Sigsbee deepsea sounding machine, invented by Commander Sigsbee of the United States Navy, allowed water depth measurements to the deepest parts of the ocean. Water depths obtained in this manner, however, were both tedious and time-consuming. By 1895 only a hundred deep soundings, in over 5000 m, had been acquired in this manner.

The first measurement of the velocity of sound in water had been obtained by Colladon and Sturm on Lake Geneva, Switzerland in 1826 (2). The experiment consisted of flashing a light and ringing a submerged bell some ten miles from an observation vessel and determining the travel time difference between the flash of light and the arrival of the acoustic signal. The result of 1430 m per second was surprisingly accurate in view of the technique and the currently accepted value of 1448 m per second at standard pressure and temperature for fresh water (3).

Shortly following the turn of the century, underwater bells were placed on a few select shoals and lightships off the New England coast, to provide warnings during periods of poor visibility. Ships required special underwater listening devices, which consisted of a submerged diaphragm-covered listening horn. It was such a mechanical system that was proposed by the Norwegian engineer M. Berggraf for obtaining water depths in 1905 (4). The development of this *bathymetre*, however, was hampered by the lack of a precise timing method.

ECHO SOUNDERS

Following the sinking of the *S.S. Titanic* on her maiden voyage in 1912, the International Ice Patrol was formed for the purpose of locating and tracking icebergs within the North Atlantic shipping routes. In the pursuit of better methods of identifying and locating icebergs, electro acoustical devices were developed. Research and development, spearheaded by the Underwater Signal Corporation (now Raytheon), produced the first electric oscillator. The system developed by Professor Reginald A. Fassenden (fig. 1.2) operated at a frequency of 540 Hz in order to generate and transmit sound pulses in the water column.

The principal purpose of the system was to provide an echo range or distance determination on icebergs. The initial sea trials took place aboard the U.S. Coast Guard Cutter *Miami* (fig. 1.3) and established that echoes from the device could be detected at ranges up to 50 miles. It was during these trials in 1914 that interfering echoes from the sea floor were observed in 31 fathoms (57 m) of water on the Grand Banks, and the echo sounder was thus discovered. Further refinements were made during the course of World War I for the purpose of submarine detection. In 1924 the first commercial fathometer, built by the Underwater Signal Company, was installed aboard the *S.S. Berkshire* on a run down the Eastern seaboard of the United States. Water depths were indicated by a flashing neon light on a calibrated dial.

4

FIGURE 1.2 Professor Fassenden, aboard the U.S.C.G. cutter *Miami*. (Courtesy of Raytheon Company.)

The year 1924 also coincides with the first discovery on land of petroleum by geophysical reflection prospecting. By the late 1920s the seismic reflection technique was being employed for petroleum exploration within the shallow bayous of the Louisiana Gulf Coast. It was not, however, until the end of World War II that the technique was applied to the open waters of the Gulf of Mexico.

FIGURE 1.3 U.S.C.G. cutter *Miami*. (Courtesy of Raytheon Company.)

HIGH-RESOLUTION PROFILERS

During the early 1950s the first high-resolution seismic profiling system, dubbed the *marine sonoprobe*, was developed at the Magnolia Petroleum Company's research laboratory. The system was based upon a magneto-strictive transducer (similar, although larger, to that built by Professor Fassenden) which operated at a frequency of 3.8 kHz. The *marine sonoprobe* produced a continuous profile of the sea floor and underlying strata to a depth of approximately 15 m (5). Although the system was originally developed for the exploration of hydrocarbons, it soon became used for geoengineering surveys along pipeline routes on the continental shelf of the northern Gulf of Mexico.

By the late 1950s the sonoprobe had been replaced in the exploration for hydrocarbons by more powerful and deeper penetrating sparkers developed at the Woods Hole Oceanographic Institute (6). Low-powered sparkers, on the order of 4 to 15 kJ, are still employed today for high-resolution geophysical surveys.

SIDE SCAN SONARS

Side scan sonar systems were developed during the late 1940s (7) as a spinoff of antisubmarine warfare sonar systems. They eventually found

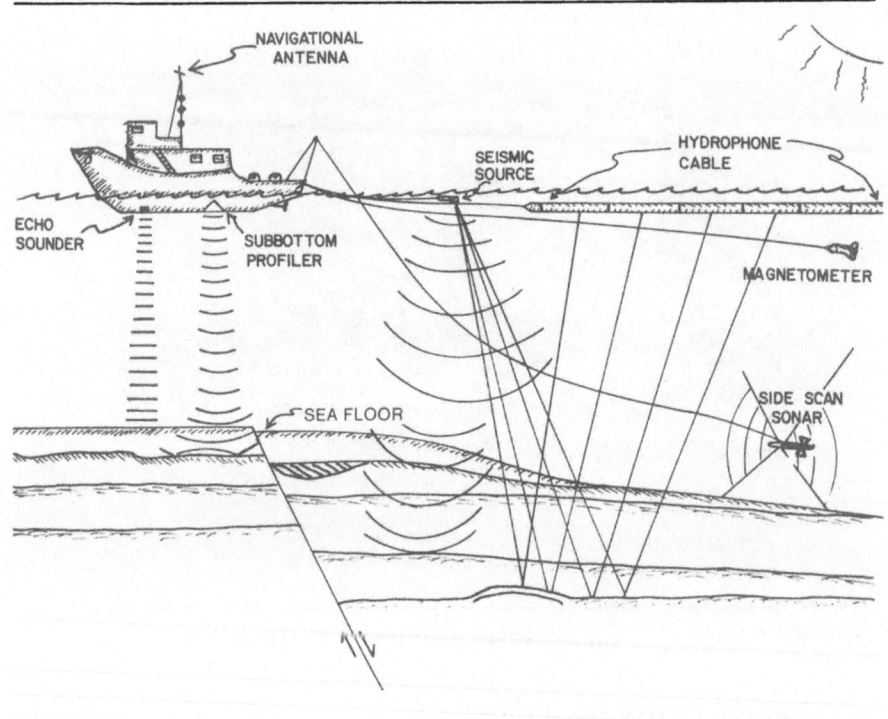

FIGURE 1.4 Diagram of sensors deployed from survey vessel. (Drawing by the author.)

application within high-resolution geoengineering surveys as an ideal tool for assessing the surface morphology of the sea floor between survey lines. Initially such systems were referred to by the acronym ASDIC, for Acoustic Sound Detection, and later as SONAR, for Sonic Navigation and Ranging.

MAGNETOMETERS
The proton precession magnetometer found its first application in high-resolution surveys in the early 1960s identifying and locating ferro-magnetic objects, such as sunken vessels and pipelines, on the sea floor.

GEOENGINEERING SURVEYS

By the late 1960s high-resolution geoengineering surveys, using up to half a dozen different tools or sensors together, were being conducted on a routine basis to assess the geomorphology of the sea floor and subbottom stratification along pipeline routes, at drilling platform sites, and at other locations where construction operations were anticipated. Figure 1.4 illustrates the deployment of high-resolution seismic sensors from a vessel during the course of a typical survey operation.

In 1972 a multifold seismic reflection program was conducted by Aquatronics (now Fairfield Industries) to collect deeper-penetrating seismic reflection data to depths of several thousand feet beneath the sea floor. As with most of the seismic survey techniques employed in geoengineering hazard studies, this multifold digital recording technique had been initially developed by the petroleum industry for the exploration of hydrocarbons. It was simply a modification of the equipment to sample the higher acoustic frequencies required for the definition of geoengineering hazards within the uppermost sediments beneath the sea floor (fig. 1.5).

TYPICAL SURVEY

In the 1980s an offshore high-resolution seismic survey requires the mobilization of a sturdy seagoing vessel (fig. 1.6) with a large number of geophysical sensing devices. The vessel and sensors must be tailored to the local environment as well as the geoengineering objectives of the survey.

Locations for the survey may vary from shallow lakes and bays to midocean operations miles from land and shore support facilities. The specific objectives may simply be to assess the nature of the sea floor and underlying strata to depths of a few meters, as in the case of jack-up drilling rig installation or pipe-laying operations, or to require a detailed investigation to sediment depths of over 1000 m in water depths exceeding this figure for production and drilling platform installations.

For each of these requirements the appropriate sensor system must be selected according to its acoustic frequency, power output, and desired resolution. Table 1.1 lists a number of sensors and general specifications, as well as optimum penetration and application toward high-resolution geoengineering surveys.

Besides the above factors, the geotechnical and resulting geophysical nature of the sea floor must be taken into account. Hard bottoms such as

8

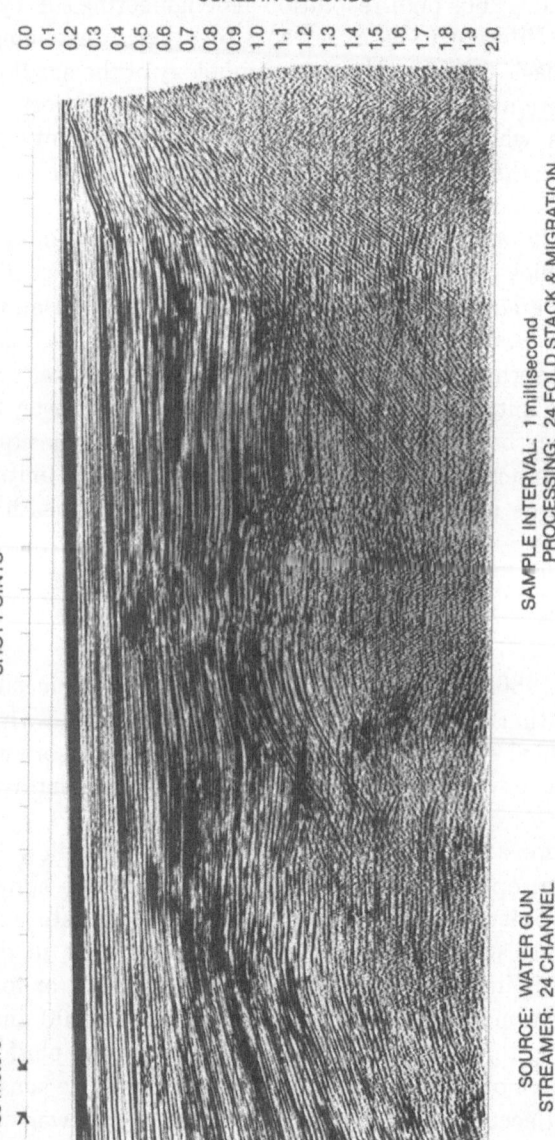

SCALE IN SECONDS

0.0 0.1 0.2 0.3 0.4 0.5 0.6 0.7 0.8 0.9 1.0 1.1 1.2 1.3 1.4 1.5 1.6 1.7 1.8 1.9 2.0

SHOT POINTS

83 meters

SOURCE: WATER GUN
STREAMER: 24 CHANNEL

SAMPLE INTERVAL: 1 millisecond
PROCESSING: 24 FOLD STACK & MIGRATION

FIGURE 1.5 Example of multifold processed seismic reflection section, offshore California. (Courtesy of Nekton, Inc.)

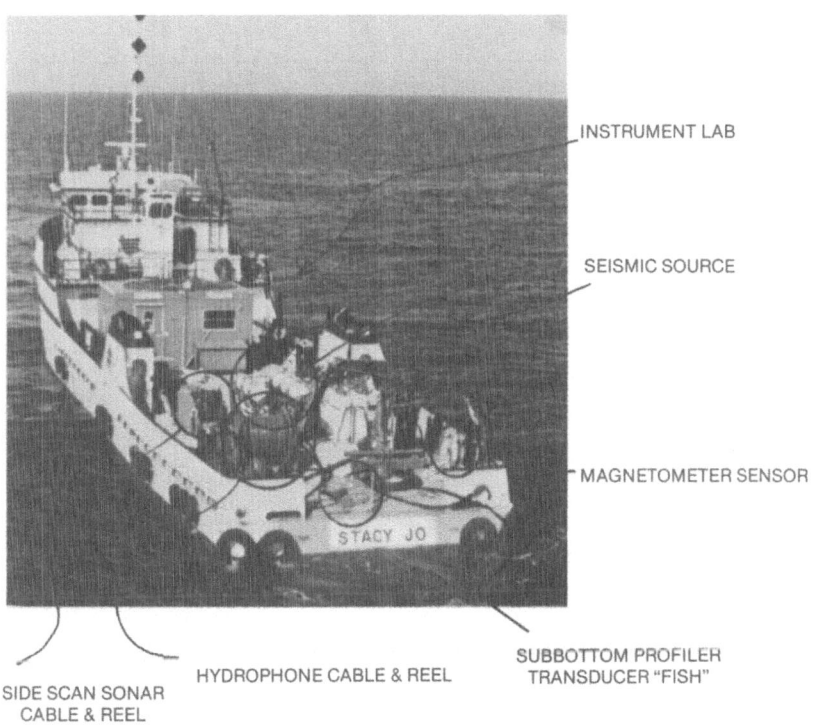

INSTRUMENT LAB

SEISMIC SOURCE

MAGNETOMETER SENSOR

SUBBOTTOM PROFILER
TRANSDUCER "FISH"

HYDROPHONE CABLE & REEL

SIDE SCAN SONAR
CABLE & REEL

FIGURE 1.6 Picture of high-resolution geophysical survey vessel. (Courtesy of Offshore Oil Services, Inc.).

limestone reefs require special sensor consideration compared to those needed for soft sediments such as found in deep waters or on deltas. Additional requirements may be imposed by the presence of sands or shallow gas-bearing sediments, as well as governmental regulations or insurance coverage that may specify the conduct of survey operations.

TEXT ORGANIZATION

The following chapters provide details as to the types of high-resolution equipment and their respective limitations according to their acoustic frequency and use in obtaining geoengineering survey data. Chapter 2 provides an overview of basic geophysical concepts as applied to the acquisition and interpretation of high-resolution geophysical data.

TABLE 1.1 High-resolution geophysical sensors:
General characteristics and applications

Sensor System	Frequency Range (kHz)	Typical Penetration (meters)	Applications
Echo Sounders	10–200	0–1	Water depths; gas bubbles in water column
Side Scan Sonars	100–200 10–100	— —	Seafloor morphology irregularities and debris
Subbottom Profilers	1.0–10	10–30	Very shallow stratigraphy, gas bubbles in water column, relict channels and gas bearing surface sediments
Seismic Reflection Systems	50–250 Hz	1000	Seismic stratigraphy, detection of faults, and bright spots
Magnetometers	—	—	Detection of ferromagnetic materials on sea floor

Chapter 3 gives a brief explanation of engineering terms employed in the field of marine geotechnique, and their applicability and relationship to the acoustic nature of marine sediments and their suitability as foundation material for offshore structures.

Chapters 4 through 10 cover respective equipment types and operations, while the final chapters summarize the value and limitations of high-resolution survey data through the interpretation process.

The subject of navigation for the positioning of a vessel during the course of geoengineering surveys is briefly covered in chapter 12, while a table of useful conversion factors in the appendix and a glossary are provided.

Two
MARINE GEOPHYSICS: AN OVERVIEW

The geophysicist uses mathematics to formulate the physical precepts involved in the transmission of sound as applied to high-resolution geophysical (HRG) surveys. The purpose of the following paragraphs is to expose the nongeophysicist to a few of those principles. Standard textbooks on the subjects of geophysics (1, 2), acoustics (3), or optics should be consulted for greater detail and the mathematical derivations.

PROPAGATION OF SEISMIC WAVES

Upon the detonation of a sound source within the water column, acoustic energy in the form of sound waves is released and travels outwards in a spherical fashion, just as wavelets propagate away from a stone tossed into a pond or radiate from a stereo speaker. The fundamental principles involved were first formulated by Christian Huygens in the seventeenth century.

The seismic source and characteristics of the media of propagation (i.e., water type and depth) determine the basic character of the seismic waves generated. These waves have amplitude (strength), frequency, and velocity (speed).

The relation between frequency and wavelength is given by:

$$\lambda = \frac{v}{f} \tag{2.1}$$

where λ is the wavelength (meters), v is the velocity (meters per second), and f is the frequency (Hertz) of a propagating acoustic wave. It is the wavelength that determines the resolution of a particular seismic system.

TYPES OF SEISMIC WAVES

There exist two types of seismic waves involved in geophysical surveying: compressional or primary P waves, and shear or S waves.

COMPRESSIONAL WAVES

The P waves move away from the source and reflectors, by alternatingly compressing and dilating the elastic media (water or sediments) along the travel path. The resultant equation describing this action is given by:

$$Vp = \frac{\sqrt{l + 2\mu}}{\rho} \qquad (2.2)$$

where Vp is the compressional wave velocity, l is a Lamé coefficient, μ is the rigidity modulus, and ρ is the density of the media through which the wave is traveling. The rigidity modulus and Lamé coefficient are proportionality constants relating shear and stress strain for ideal homogenous and continuous elastic solids (1). Compressional waves are the type involved in the HRG reflection-profiling techniques discussed herein.

SHEAR WAVES
The shear or S waves oscillate in a direction normal to (at 90°) the travel axis, and are described mathematically by:

$$Vs = \sqrt{\frac{\mu}{\rho}} \qquad (2.3)$$

where Vs is the velocity of shear waves, ρ is the density, and μ is the rigidity modulus of the media through which they are traveling. Shear wave velocities are lower than those of P waves. Due to the lack of elasticity of water, however, shear waves do not propagate through the water column, and are not generally used for HRG profiling methods.

WAVE PATH THEORY
The mathematical details describing three-dimensional elastic wave propagation are complex, and are more simply understood for our purposes by wave path theory (1). This theory is described by Huygen's Principle, whereby every point on a wave front is the source of a new wave that also travels out from it in a spherical manner. If the spherical waves have a large radius they can be treated as planes, and lines perpendicular to the wave fronts, called wave paths or rays, can be used to describe the propagation of seismic waves. If one assumes a two-dimensional model with an overlying water column and subbottom strata, one may trace the propagation of seismic waves by this simple theory involving straight lines and connecting successive wave front positions as shown in figure 2.1. As illustrated, the angle of incidence is equal to the angle of reflection. In the

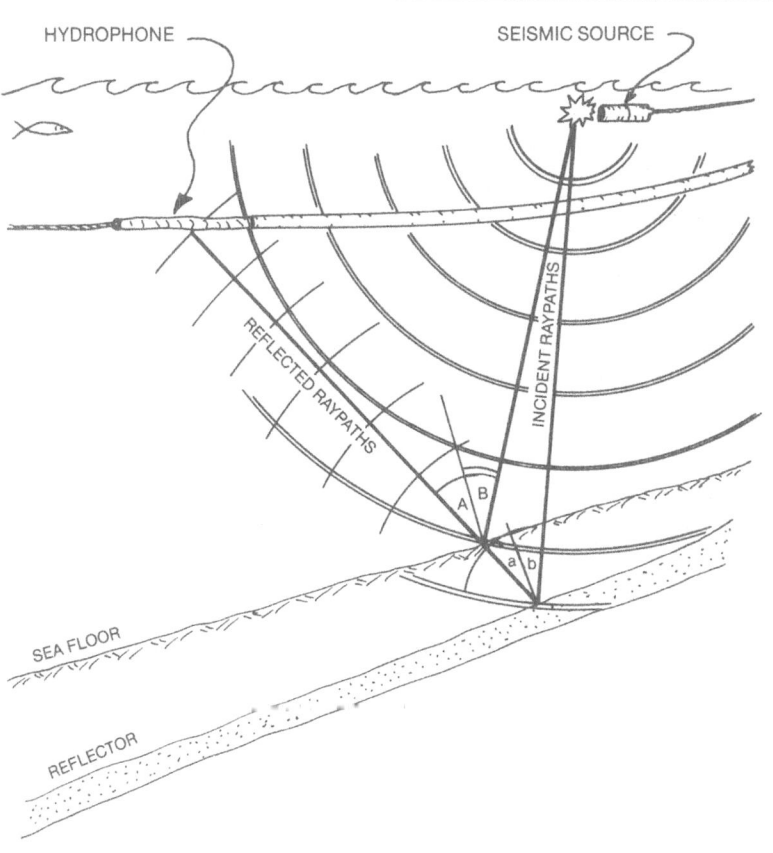

FIGURE 2.1 Two-dimensional seismic reflection model with raypaths. (Angle of incidence B = angle of reflection A.) (Drawing by the author.)

recording of continuous seismic profiles it is the shortest travel path that is recorded (first arrival reflection), the path of which is represented by the normal angle of incidence on the reflecting horizon.

While wave path theory simplifies the understanding and mathematical formulation of the propagation of seismic waves, it does not consider the actual three-dimensional spatial geometry frequently encountered on and beneath the sea floor. Thus, the interpretation of HRG seismic data from steeply dipping areas, as along the continental slope, needs to account for such factors.

REFLECTIONS

Seismic waves are reflected at interfaces, such as the sea floor and stratigraphic or lithologic and geotechnical boundaries, due to changes in the product of density and velocity (acoustic impedance) of the layers. Thus, HRG seismic reflections recorded during a survey are obtained at well-defined stratigraphic interfaces, as well as where subtle changes in physical properties occur.

The amount of reflected energy is a function of the thickness of the reflecting layer, the sharpness of the strata boundary, and the signal reflectivity at a given sediment boundary or physical interface. The reflectivity is proportional to the product of the velocity and density on either side of the interface according to the equation:

$$R = \frac{V\rho(2) - V\rho(1)}{V\rho(2) + V\rho(1)} \tag{2.4}$$

where R is the reflection coefficient and the product of velocity and density for each layer and $(V\rho)$ is referred to as the acoustic impedance. Thus, the acoustic reflectivity of a seismic signal depends upon the contrast in the product of density and velocity on opposite sides of an interface boundary.

If the acoustic impedance for two layers were equal, $[V\rho(1) = V\rho(2)]$, there would be no reflection from the interface in spite of variations of both density and velocity. On the other hand, the greater the impedance contrast (i.e., the farther it is from unity), the stronger the reflector that will be produced.

It should be noted that if an underlying layer is of lower impedance than the overlying layer, the reflected wave is phase inverted by 180°. This is an important criteria in the identification of gas-bearing low-velocity sediment layers, as the resultant bright spot is of reversed phase or polarity as shown in figure 2.2. Such phase reversals also occur for reflections from the sea surface to air interface, which produce multiples and ghosts.

If both the velocity and reflectivity coefficient for a given reflector are known, as may be the case for multifold HRG reflection data, it should be possible to determine the density of the strata. However, this has yet to be performed in a routine manner for obtaining either the physical properties of sediments or providing a means for determining the presence of hydrocarbons.

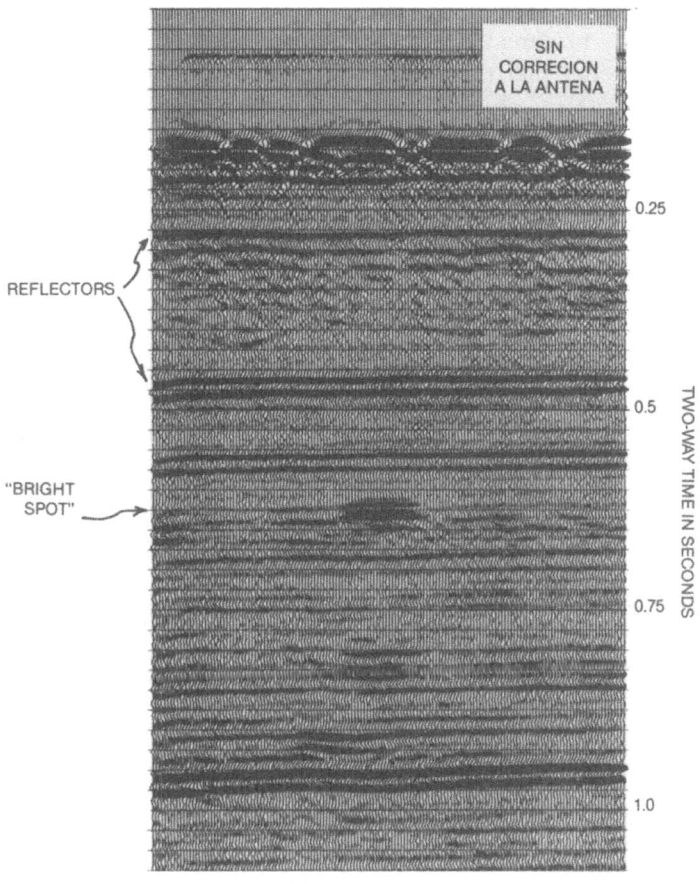

SIN
CORRECION
A LA ANTENA

0.25

REFLECTORS

"BRIGHT SPOT"

0.5

0.75

1.0

TWO-WAY TIME IN SECONDS

FIGURE 2.2 Bright spot, with phase inversion, 12 fold processed record from offshore Venezuela. (Courtesy of BEICIP/CGG/INTEVEP.)

GEOMETRY OF REFLECTION PATHS

Upon arrival at an acoustic interface, seismic waves are reflected according to wave path theory at equal angles to either side of a perpendicular line to the reflector, as illustrated in figure 2.1. Thus, if shooting a seismic profile over a flat bottom, the returned echo or reflection travels vertically from the source and back up to the receiver at an angle of 90° to the reflecting horizon. In this case both incident and reflected waves are at normal angles with respect to the reflecting interface.

If the process is carried out over a steeply sloping surface, however, the recorded reflection is produced from a point up-slope of the source/receiver position, in order that the angle of incidence and reflected wave bisect a perpendicular to the slope. In order to correct for this effect it is necessary to apply a mathematical correction referred to as *migration*. This may be performed by a computer during data processing, or manually on a point-by-point basis (1).

ABSORPTION AND ATTENUATION

The physical or elastic characteristics of the water column and solids comprising marine sediments determine the amount of seismic wave energy that will be absorbed and thus attenuated. This effect reduces the signal strength and resultant amplitude of the seismic reflection signals.

The *attenuation* is a function of both the spherical propagation of the signal (a function of the cube root of the distance traveled) and the acoustic *absorption* of the media through which the signal travels. Thus, recorded seismic profiles for which these effects have not been compensated show a decrease in signal amplitude with time (depth) of recording. In practice, this loss in amplitude is compensated for, either in real time through the application of time-varying electronic gain amplification (TVG), or during postsurvey data processing.

WIDE-ANGLE REFLECTION AND REFRACTION

By separating the source and receiver of a given high-resolution seismic system one obtains progressively wider reflection angles and eventually refractions. The refraction and wide-angle reflection techniques are excellent methods for obtaining subbottom velocity information, as they are easily implemented with single channel HRG seismic reflecting tools.

When the distance between the source and receiver reaches a certain critical separation, seismic waves are found to refract or bend and travel along the underlying submarine strata, as illustrated in figure 2.3. This refraction of seismic waves occurs when the angle of incidence (fig. 2.3) reaches a certain critical value governed by Snell's Law (1). This critical angle at which refraction occurs may be computed by:

$$i = \sin^{-1}(V1/V2), \tag{2.5}$$

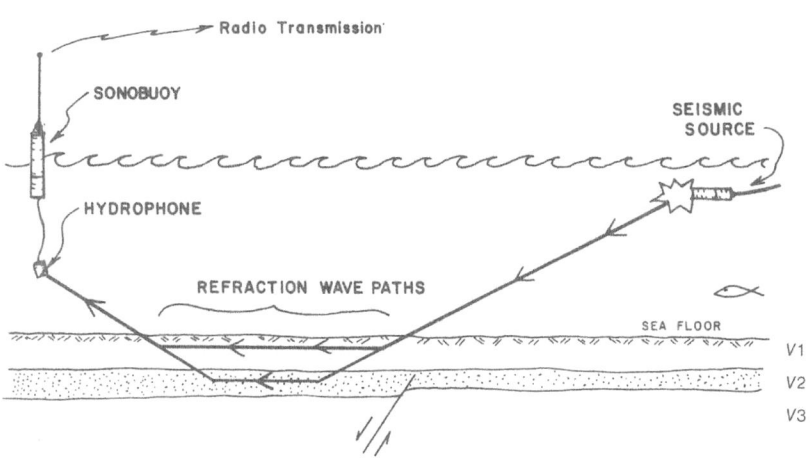

FIGURE 2.3 Two-dimensional model of raypath during seismic refraction. (Drawing by the author.)

where i is the critical angle of refraction in degrees, and $V1$ and $V2$ are the respective compressional acoustic velocities for the layers above ($V1$) and below ($V2$) the interface (fig. 2.3).

As the time involved during refraction of a seismic wave is directly proportional to the velocity of the strata through which the refraction wave is traveling, the velocity for a given layer may be obtained from a time-distance plot (fig. 2.4).

Sonobuoys may be deployed for obtaining wide-angle seismic reflections and refraction profiles at sea (fig. 2.4). Such buoys consist of a hydrophone and radio transmitter that relay the received seismic signals to the survey vessel, where they are received by radio and recorded graphically, as the survey vessel travels away from the buoy's deployment position while firing its seismic source. The delay in arrival times for a given reflector, as the source and receiver are progressively separated, is referred to as *normal moveout* (NMO). The delay that results from arrival times at two separate hydrophones is referred to as *moveout*. It is this delay or moveout that differentiates between reflections and interference (i.e., multiples) on seismic reflection records. The NMO delay is proportional to the average velocity down to the reflecting horizon. The NMO technique is used to obtain velocity information from multifold reflection profiles, where a long string of individual hydrophone groups is towed behind the survey vessel (chap. 9).

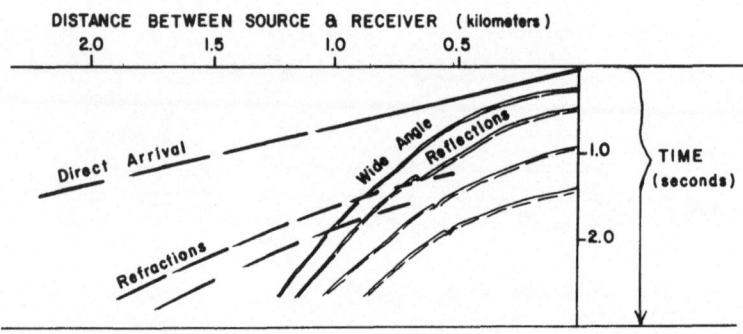

FIGURE 2.4 Time-distance plot typical of wide-angle seismic reflection, and refraction recording. (Drawing by the author.)

It should be noted that velocities acquired from refraction data are not necessarily the same as those obtained from waves traveling vertically (4), due to the acoustic anisotropy of sediment layers; however, sedimentary rocks are usually considered isotropic mediums.

SEISMIC VELOCITIES

As shown by equation (2.2), seismic waves travel through various media as a function of the physical properties. The velocity of sound in water thus varies according to the density, which in turn varies as a function of salinity, temperature, and depth (pressure) of the water. In order to obtain precise velocities of the water column at sea, it is necessary to either compute the velocity from a profile of the forementioned properties (5), measure the velocity in place with a velocimeter, or obtain an average velocity for a precisely known depth (obtained by bar check or pneumatic gauge) by measuring the travel time. The latter technique is simpler and provides the basis for the velocity/depth calibration technique used in echo sounding (chap. 4).

The same technique is applicable toward obtaining velocities within the underlying sediments, in order to compute the true depth of recorded strata. If seismic reflections are obtained through wide angles, as is the case for multifold or sonobuoy data acquisition, the average velocities may also be obtained from the NMO delays.

The curve of figure 2.5 is a compilation of numerous velocity measurements for the uppermost 1500 m of marine sediments. As can be expected, the velocity increases with respect to depth, although it is usually slightly lower than the overlying water column just below the sea floor. This low-velocity layer (LVL) or weathered layer (when exposed to

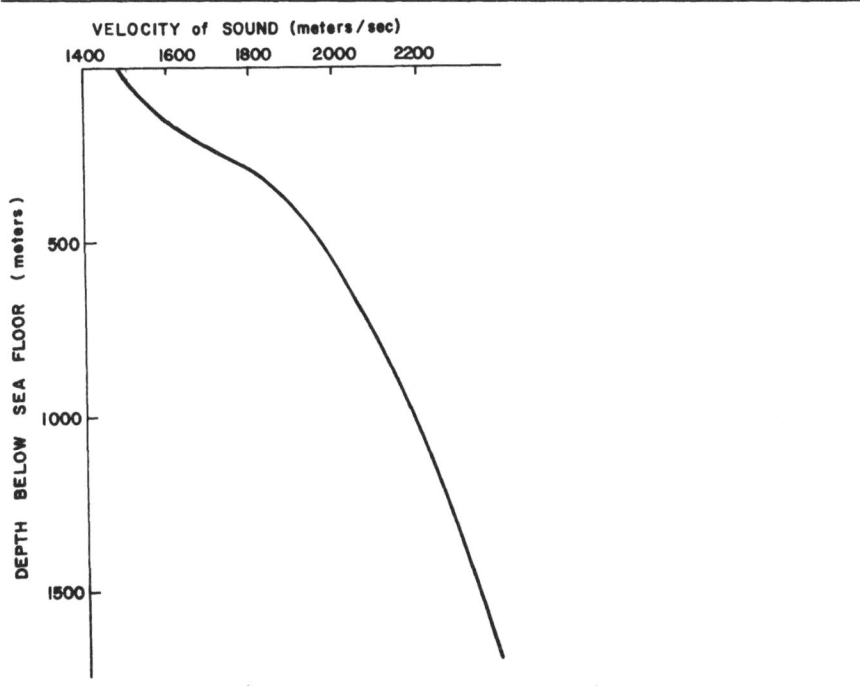

FIGURE 2.5 Typical curve of acoustic velocity variation with depth. (From Trabant, P. K., 1972, Consolidation characteristics and related geotechnical properties of sediments retrieved by the *Glomar Challenger* from the Gulf of Mexico: unpublished M.S. thesis, Texas A & M University, p. 77.)

air) is usually due to the presence of gas or air within the sediment pore spaces.

Variations in subbottom velocities, however, may be expected as a function of the specific geology and lithologic composition within a survey area. If the sea floor is composed of well-indurated carbonates (limestone), very high velocities may occur, while a thick wedge of recent deltaic sediments may have velocities considerably lower than that of water. However, the most important factor governing the velocity of sound within sediments and rocks is the granular structure and intergranular void space (porosity). As these may vary widely for a given type of sedimentary rock so will the acoustic velocities.

RESOLUTION AND FREQUENCY

Frequencies employed for HRG surveys cover a broad spectrum, ranging from about 100 Hz for deep-penetrating systems to over 200 kHz for echo sounders and side-scanning sonars.

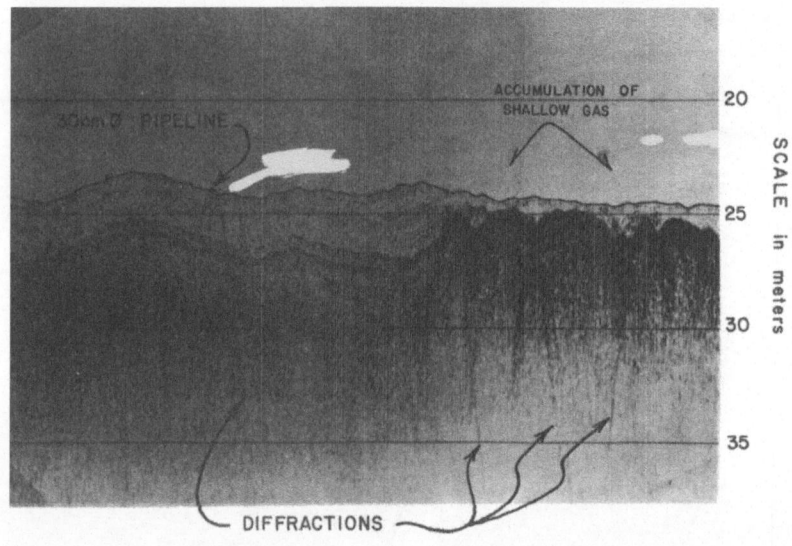

FIGURE 2.6 Typical diffraction patterns from a pipeline and shallow gas pockets as revealed on a 14-kHz subbottom profiler recording off the Mississippi River delta. (Courtesy of Ferranti–O.R.E., Falmouth, Mass.)

The resolution of a given seismic reflection system is the ability to distinguish between adjacent or overlying strata. The ability of a given system to resolve a particular unit is a function of the source frequency and the receiver/recorder system, and its ability to display a seismic reflector. While mostly a function of the frequency, other factors must also be taken into consideration, such as the thickness of the reflecting horizon and the sharpness of the boundary between the strata producing a reflection.

In theory, the resolution of a system is on the order of a quarter wavelength (λ). Thus, if the velocity of a seismic wave were 1600 m per second at a center frequency of 3.5 kHz, the $1/4$ λ or optimum resolution would be 0.12 m. In practice, however, this is rarely achieved, and resolution on the order of a third or half of λ is more common.

DIFFRACTIONS AND OTHER UNWANTED SIGNALS

In addition to reflections from submarine strata, the high-resolution seismic profiling technique produces a number of signals, due to the nature of the propagation of acoustic signals. These include: diffractions, surface *ghosts* (water surface echoes), direct arrival signal, side echoes, and multiples.

Diffractions

As seismic waves propagate in a spherical fashion through the water column and sediments, they may be reflected by certain geological or man-made features that act as a point source to produce diffraction patterns as illustrated by the recording of figure 2.6.

While diffraction signals are usually considered interference, they may yield valuable information for identifying such features as boulders, fault planes, buried pinnacle reefs, canyon and valley axes and ridges, small bright spots, pipelines, and debris in the water column as well as on or beneath the sea floor.

Water Surface Echoes or Ghosts

Water surface echoes are the result of signal reflections from the air/water interface above both the source and hydrophone receiver array. Their locations, with respect to time, are a function of the submerged depth of the source and hydrophone, and may interfere with desired reflection signals. The ghost signal may also be used to reinforce either the seismic source pulse or hydrophone receiver signals by appropriate depth spacing, corresponding to the wavelength of the frquencies involved. (See chapter 8 for nomogram and computation.)

Direct Arrival

Direct arrival is the recording of the source signal as it travels directly from source to receiver, and its arrival time is proportional to the distance between the two and the acoustic velocity of the water. Interference from the direct arrival may be reduced by appropriate spacing between adjacent hydrophones attached in parallel.

Side Echoes

The recording of side echoes is the result of reflections from the spherical spreading of the seismic source waves and from features adjacent to the survey profile line. Examples are: steep slopes and intercanyon ridges, pinnacle reefs, and other such features that may produce misleading reflectors.

Multiples

Multiples are reflections that result from the echoing of the seismic signal between the sea floor and surface (water column multiples) or between reflecting strata, or even combinations of such signal paths. Multiples are events that have undergone more than one reflective cycle. The occurrence of multiples is one of the most deleterious effects in interpretation since it masks desired reflecting horizons. Removal of multiples is one of the primary purposes of the multifold seismic technique.

Sea-floor multiples are identified by their timing, which is equivalent to the water depth. Thus, if the sea floor occurs at a time of 100 msec, the first multiple will appear at a time of 200 msec (reversed in phase), at 300 msec for the second multiple, and so on, as shown in the recording of figure 2.7.

ENHANCEMENT OF SEISMIC REFLECTION DATA

Since obtaining the first continuous seismic profiles (CSP) on rotating drum recorders in wiggly-line trace format, considerable research has been devoted toward enhancing HRG seismic reflection data. This has essentially involved increasing the signal-to-noise ratio to obtain clear reproductions of subbottom stratification with the highest resolution possible.

Improvements in hardware (sources, receivers, and recorders) were first made and then followed by enhancement of analog signal processing techniques. By the early 1970s digital techniques were applied, and further improvements in data acquisition are still taking place, mostly through developments in signal processing software.

Within analog systems, signal enhancement is performed by gain amplification and the filtering of undesirable portions (nondata) of the frequency spectrum. Digital systems perform these same functions, and additional ones, by operating on the digital binary bits, representing the analog signal, within a microprocessor.

With the advent of the microprocessor and availability of dense storage media (magnetic tape), the Common Depth Point (CDP) technique was developed in the late 1960s. Reflections from a specific subbottom point, obtained with different source/receiver geometries, can be summed to produce an appreciable increase in signal-to-noise ratio.

PRINCIPLES OF SEISMIC INTERPRETATION

Recordings of seismic reflection profiles consist of alternating positive and negative wave forms depicting frequency and amplitude as a function

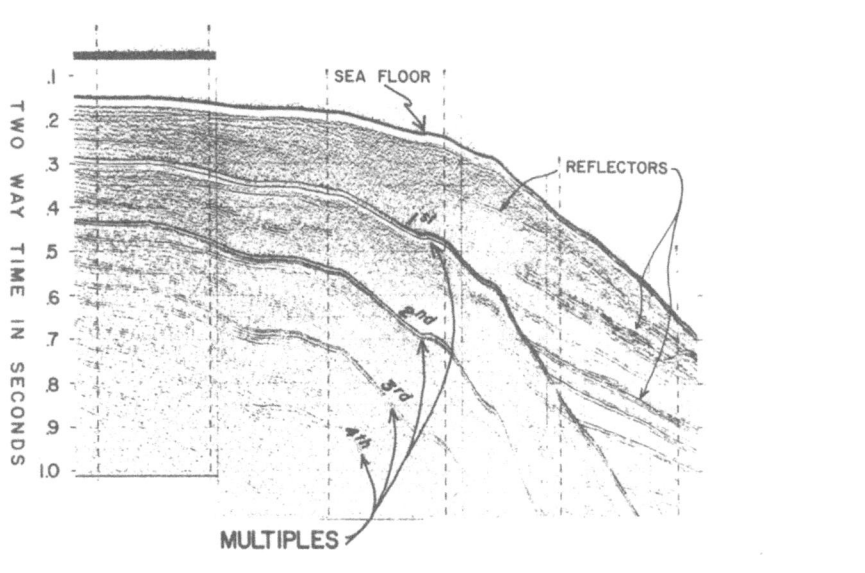

FIGURE 2.7 Illustration of multiple reflections on analog seismic recording from offshore U.S. East Coast. (Courtesy of Summit Geophysical International, Inc.)

of time and distance. The amplitude and frequency of the recordings are the result of the amplification and filtering of the reflected signals. Continuous reflections are discerned by virtue of similarities in wave form, which are generally related to subbottom stratigraphy.

While HRG seismic recordings may first appear to depict the subbottom stratigraphy, it should not be overlooked that they in fact represent time, a function of the velocity of the traveling acoustic signal, and the acoustic reflectivity of the strata, which in turn are functions of the stratigraphy and lithology.

The horizontal scale is a function of the survey vessel's speed, source firing rate, and the recorder sweep-rate. The ratio of true horizontal to vertical scale is referred to as *vertical exaggeration*, which may be very large (values on the order of 20 or 30 are not uncommon), and produce a distorted picture of the sea floor and underlying stratigraphy.

CORRELATION OF REFLECTORS

Interpretation consists of the identification of prominent reflectors and their correlation throughout a survey area in order to produce a map of the geologic structure. Visual correlation is the main technique, but processed data can be produced whereupon correlations of signal according to density (amplitude), velocity, or frequency may be presented in color to facilitate interpretation.

When subsurface data are available in the form of drill cuttings or detailed analysis of geotechnical properties from engineering boreholes, the interpretation process is greatly enhanced.

GEOENGINEERING HAZARDS

Hazards to the emplacement of drilling or other structures on and beneath the sea floor involve the suitability of the subbottom as a foundation, and the absence of structural (geologic), stratigraphic and other features that may impede their emplacement or operations. Typical hazards include: lateral variations of physical properties (such as occur along buried river beds), faults, and overpressured (relative to hydrostatic) gas-bearing sediment pockets, which may cause a blowout during drilling operations.

Three
MARINE GEOTECHNIQUE, AN OVERVIEW

The study of the geotechnical properties of marine sediments has evolved over the past several decades as a natural expansion of the study of soil mechanics. The field involves the "study of the physical properties of marine sediments and the response of a sedimentary system to applied static and dynamic forces" (1).

The following pages are intended to give the reader a brief and general description of the subject. It is not, however, intended to be a detailed review, such as may be found within basic texts on the subject of soil mechanics and foundation design (2–6).

SEDIMENT-STRUCTURE INTERRELATIONS

Marine sediments react to the emplacement of structures such as platform legs or pipelines in what is known as soil-structure interaction. Based upon laboratory or in-situ test results, the civil engineer attempts to predict the behavior of soils and of the structure they support. For example, he or she may determine the amount of settlement that will take place following the emplacement of a structure on the sea floor, or assess the necessary depth to which piles should be driven to support a platform subjected to a designed maximum load (e.g., drilling rig superstructures, processing facilities, storm wave forces, winds, etc.).

DESIGN CONSIDERATIONS

The legs of a jack-up drilling rig may settle up to several meters into the sea floor, over a predetermined period of time, depending upon the geotechnical properties of the underlying sediments, and load imposed by the structure (5). It is the unforeseen settlement beyond the initial spud-in or preloading that is of concern. A drilling structure may similarly require the insertion of piles to depths of over 100 m to support the necessary drilling equipment and provide a secure foundation to resist a severe environment, such as hurricanes, earthquakes, or unstable slopes on the sea floor.

An improper evaluation of the geotechnical properties may result in the failure of a jack-up drilling rig by puncture through a thin hard layer overlying softer soils, or by the differential settlement of the support legs caused by horizonal variations of the geotechnical properties (7). The appropriate depth of insertion of deep piles for a platform or wellhead is critical with respect to their cost, as an overdesigned depth requires deeper insertion and a proportional increase in cost.

BASIS FOR SEDIMENT SAMPLING

Ideally, the selection of an appropriate borehole site for obtaining core samples for geotechnial analyses should be predicated upon the local geology, which is determined from the interpretation of a high-resolution geoengineering survey. The detailed review and correlation, between the survey data and geotechnical test results, provides the necessary information for the establishment of a sound foundation design within the offshore environment. More frequently, however, borehole locations are selected on the basis of exploration objectives without the benefit of HRG data.

SAMPLING METHODS

A typical offshore boring is obtained from an anchored vessel using a rotary drilling rig, such as shown in figure 3.1. Deeper water (greater than about 300 m) operations require dynamically positioned drilling vessels, which do not employ anchors. Samples are obtained as drilling proceeds to typical depths of 100 m beneath the sea floor using a wireline sampler, operated through the bore of the drillpipe.

In place or in-situ testing may also be performed down the drillstem during such drilling operations. Gravity driven (free fall) cores, up to ten m in length, may also be used for sampling near-surface sediments to shallow depths as in the case of surveys along pipeline routes.

SEDIMENT SAMPLES

The sediment samples obtained by the above coring techniques should be handled with care to avoid being physically disturbed. Samples are sent to a shore-based geotechnical laboratory for testing and analyses. Proper storage and labeling of such samples are critical for their identification in the laboratory. Core samples should be maintained as close to their in-place orientation (e.g., vertical) and environment (cool and moist) as possible. Labels on the samples should indicate the sample orientation (top, bottom), location, and other pertinent facts.

GEOTECHNICAL ANALYSES

The results of the tests are usually presented as a graph of the various physical properties plotted with respect to depth, such as shown in figure 3.2. The basic battery of laboratory tests performed upon marine sediment samples in evaluating their geotechnical properties may include:

FIGURE 3.1 Vessel engaged in drilling geoengineering borehole. (Courtesy of McClelland Engineers.)

grain size analyses, water content, Atterberg limits, bulk density, and shear strength. Further testing may also include: specific gravity of the solids, porosity or void ratio, mineralogic composition, pore water chemistry, and consolidation properties.

The following are brief descriptions of the tests and analyses that may be performed on core samples obtained for geotechnical engineering purposes.

SOIL OR SEDIMENT "INDEX PROPERTIES"

Visual Description
Visual description of the sediments retrieved is done subjectively, sometimes with the aid of a color guide chart (8). For example, a typical

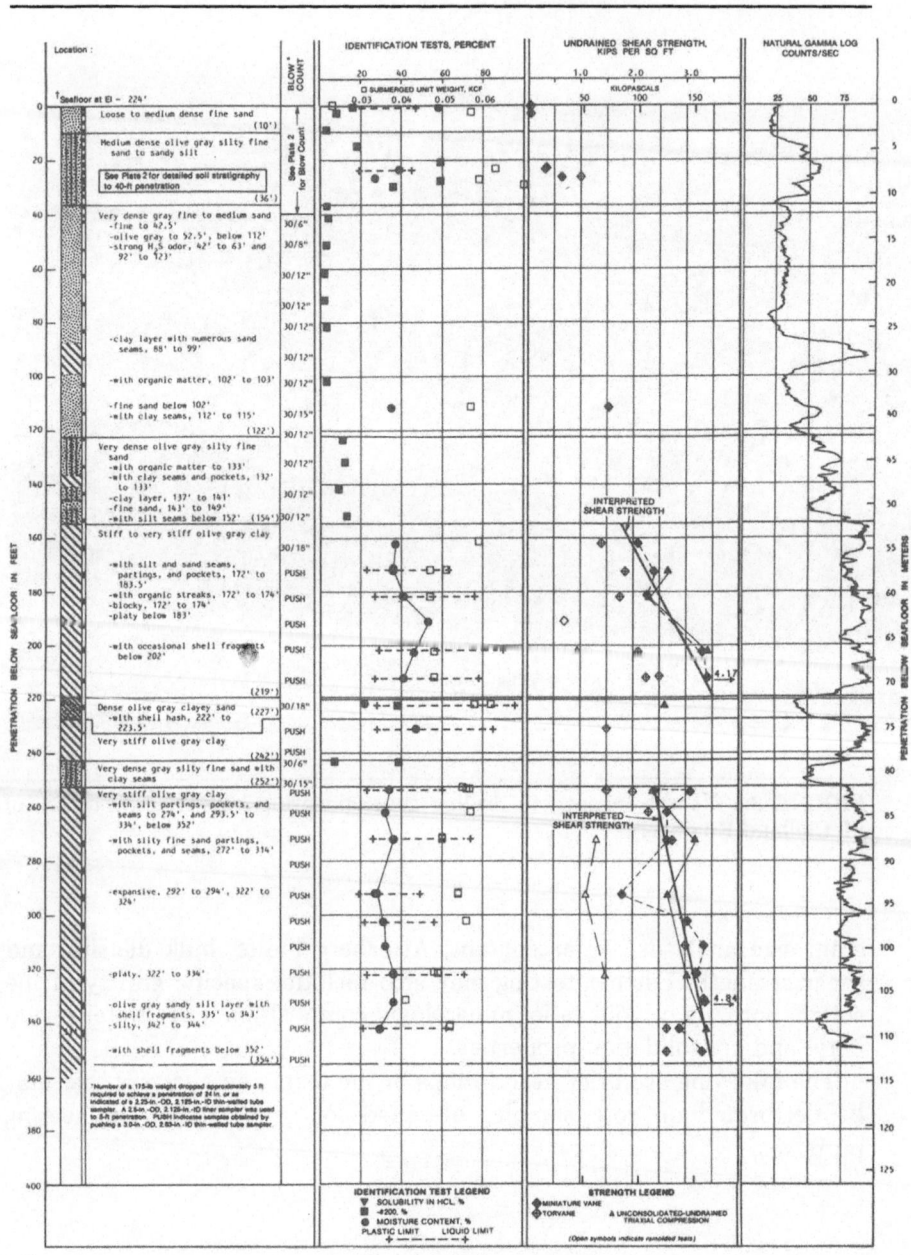

FIGURE 3.2 Typical geoengineering profile of borehole data. (Courtesy of McClelland Engineers.)

description could be: dark brown or red silt, containing fragments of shell and organic debris and thin (1 mm) sand laminations.

The relative strength may also be described in the field on the basis of measurements made by a miniature vane, torvane, or pocket penetrometer.

Odor

The *odor,* when present, should equally be noted in a preliminary shipboard description, as it may be an indication of biogenic activity (i.e., the odor of hydrogen sulphide or rotten eggs is indicative of the decay of organics) and presence of gas within the sampled sediments.

Mineralogic Composition

The *mineralogic composition* of sediment particles may be determined visually by a perceptive geologist, but usually requires observation under a microscope with chemical or mass spectrometer analysis for accurate results. Typical compositions may include: clays (aluminosilicates, the product of weathering), carbonates (i.e., coral reef rubble), silica (beach sands), and organics (peat).

While such detailed analyses are not performed routinely as part of a geotechnical program, the composition may be critical to an assessment of the sediment as a foundation. For example, the geotechnical properties are measurably different between sands of carbonate and silica origin, in spite of their similar texture.

Grain Size Analyses

Grain size analyses permit classification of a sediment sample according to the percentage of particles within standard size categories of clay, silt, sand, and gravel as shown in figure 3.3. Two methods are commonly used to determine the relative grain size proportions or *texture*. The sieve method is used for particles greater than about 80 μm in diameter (fine sands), and a settling and sampling technique or the hydrometer method employed for finer grained particles (silts and clays). The use of electronic particle counters, although more expensive and complex, has provided a more rapid means of performing size analyses on fine-grained marine sediment samples.

A sediment composed chiefly of clay size particles, with lesser amounts of silts and sands, is referred to as a sandy silty clay. Other classification schemes for marine sediments have been presented by Shepard (9), and entire texts (10) go into the details of clay mineralogy.

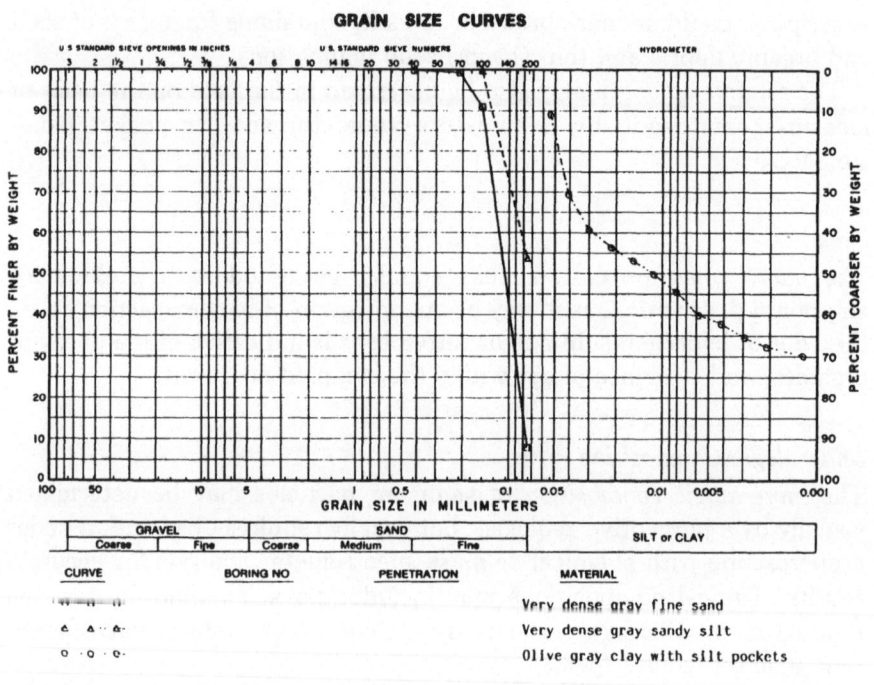

FIGURE 3.3 Graph of particle size distribution analysis, for typical marine clay. (Courtesy of McClelland Engineers.)

Sediments or soils may be classified as to texture by one of several systems according to the *grain size distribution*. Marine sediments are usually categorized according to the Unified Soil Classification System (table 3.1) established by the U.S. Army Corps of Engineers and the U.S. Bureau of Reclamation. Several other systems are occasionally used in the United States and in other countries; however, those standards set by the American Society for Testing and Materials are generally used as the standard.

Bulk Physical Properties
Bulk physical properties of a sediment sample are a measure of the relative proportions of solids, fluids, and gases. The *bulk density* is the weight-to-volume ratio of a sediment, and is expressed in kilograms per cubic meter (pounds per cubic foot). If the moist sediment sample (passing the No. 40 sieve) is often dried at 105°C for a period of 24 hours (13), then weighed and the volume assessed, one may determine the water content of the sample (now dried) by the difference in weight and volume to obtain the void ratio and porosity (see fig. 3.4 and 3.5).

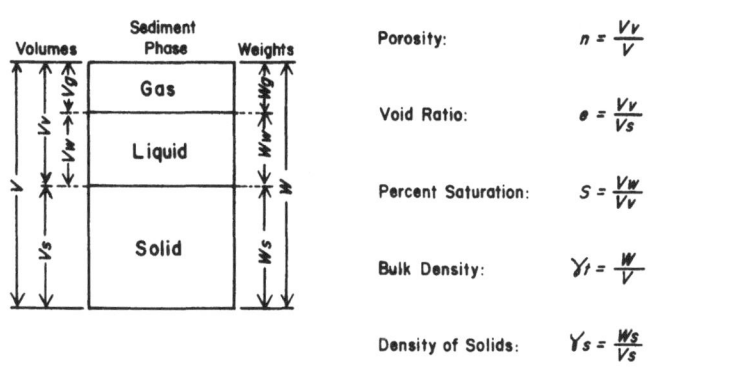

FIGURE 3.4 Block diagram of voids/solids components for a marine sediment sample, with basic formulas and equations for the computation of bulk physical properties.

Atterberg Limits

The *Atterberg limits* represent a relative measure of the liquid, plastic, and solid state of a sediment sample and are based on a set of empirically concocted tests for each of which the water content is obtained (fig. 3.6). The values are expressed as water contents (in percentage) of the sediment at points when the sample behavior shifts from that of a solid to a plastic (plastic limit), and from that of a plastic to a liquid (liquid limit). For example, if a sediment's liquid limit value was determined to be 100%, and the in-place value of the water content was 110%, the sediments would be in a liquid state. Such a state is normally not very conducive to the establishment of a solid foundation, unless, as for a pipeline, the structure were designed to literally float upon the sediments. Past experience over several decades has shown the Atterberg limits to provide a valuable understanding of soils in terms of their ability to act as a foundation material. The tests are not applicable to sediments containing grains coarser than about 40 μm, such as fine sands. The above properties (description, texture, bulk properties, and Atterberg limits) are collectively referred to as index properties.

STRENGTH OF SEDIMENTS

Shear Strength

The *shear strength* of a sediment or soil is defined as the resistance to sliding within a soil mass. This shearing resistance is a combination of

TABLE 3.1

Determine percentages of sand and gravel from grain-size curve.
Depending on percentage of fines (fraction smaller than No. 200 sieve size), coarse-grained soils are classified as follows:
Less than 5 per cent............................. GW, GP, SW, SP
More than 12 per cent............................ GM, GC, SM, SC
5 to 12 per cent............................. Borderline cases requiring dual symbols**

Major divisions			Group symbols	Typical names	Laboratory classification criteria
Coarse-grained soils (More than half of material is larger than No. 200 sieve size)	**Gravels** (More than half of coarse fraction is larger than No. 4 sieve size)	Clean gravels (Little or no fines)	GW	Well-graded gravels, gravel-sand mixtures, little or no fines	$C_u = \dfrac{D_{60}}{D_{10}}$ greater than 4; $C_c = \dfrac{(D_{30})^2}{D_{10} \times D_{60}}$ between 1 and 3
			GP	Poorly graded gravels, gravel-sand mixtures, little or no fines	Not meeting all gradation requirements for GW
		Gravels with fines (Appreciable amount of fines)	GM* (d, u)	Silty gravels, gravel-sand-silt mixtures	Atterburg limits below "A" line or P.I. less than 4 / Above "A" line with P.I. between 4 and 7 are borderline cases requiring use of dual symbols
			GC	Clayey gravels, gravel-sand-clay mixtures	Atterburg limits above "A" line with P.I. greater than 7
	Sands (More than half of coarse fraction is smaller than No. 4 sieve size)	Clean sands (Little or no fines)	SW	Well-graded sands, gravelly sands, little or no fines	$C_u = \dfrac{D_{60}}{D_{10}}$ greater than 6; $C_c = \dfrac{(D_{30})^2}{D_{10} \times D_{60}}$ between 1 and 3
			SP	Poorly graded sands, gravelly sands, little or no fines	Not meeting all gradation requirements for SW
		Sands with fines (Appreciable amount of fines)	SM* (d, u)	Silty sands, sand-silt mixtures	Atterburg limits below "A" line or P.I. less than 4 / Limits plotting in hatched zone with P.I. between 4 and 7 are borderline cases requiring use of dual symbols
			SC	Clayey sands, sand-clay mixtures	Atterburg limits above "A" line with P.I. greater than 7

32

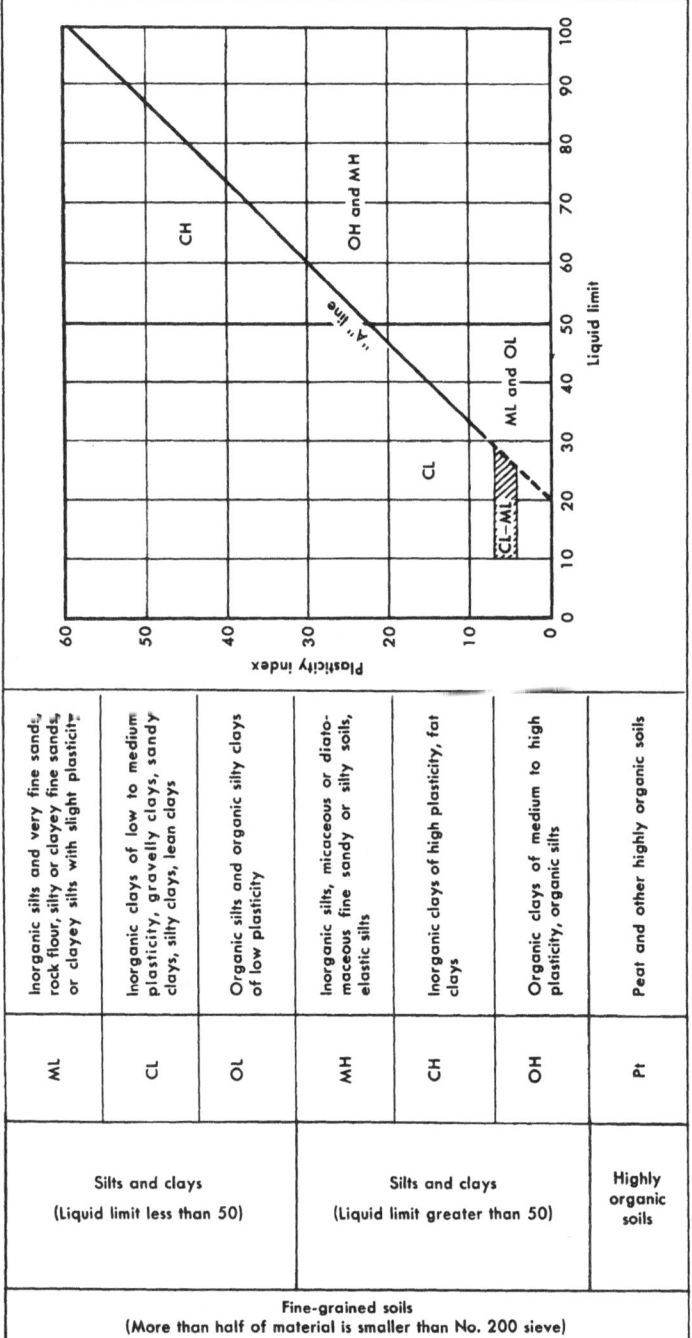

Fine-grained soils (More than half of material is smaller than No. 200 sieve)	Silts and clays (Liquid limit less than 50)	ML	Inorganic silts and very fine sands, rock flour, silty or clayey fine sands, or clayey silts with slight plasticity.	
		CL	Inorganic clays of low to medium plasticity, gravelly clays, sandy clays, silty clays, lean clays	
		OL	Organic silts and organic silty clays of low plasticity	
	Silts and clays (Liquid limit greater than 50)	MH	Inorganic silts, micaceous or diatomaceous fine sandy or silty soils, elastic silts	
		CH	Inorganic clays of high plasticity, fat clays	
		OH	Organic clays of medium to high plasticity, organic silts	
	Highly organic soils	Pt	Peat and other highly organic soils	

*Division of GM and SM groups into subdivisions of d and u are for roads and airfields only. Subdivision is based on Atterburg limits; suffix d used when L.L. is 28 or less and the P.I. is 6 or less; the suffix u used when L.L. is greater than 28.
**Borderline classifications, used for soils possessing characteristics of two groups, are designated by combinations of group symbols. For example: GW-GC, well-graded gravel-sand mixture with clay binder.

Note: Reprinted from *PCA Soil Primer* (EB007S) by permission of the Portland Cement Association, Skokie, Illinois.

33

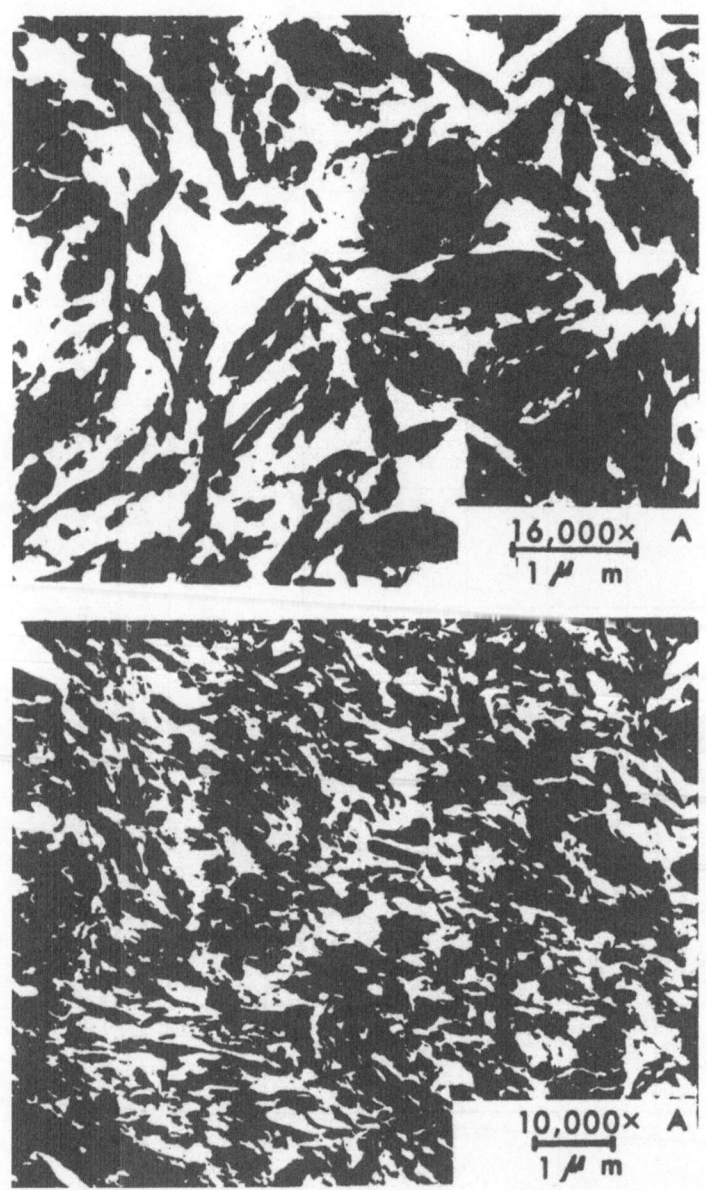

FIGURE 3.5 Electron micrographs of marine sediment (clay) at two stages of consolidation. Note: Change in intergrain void space (porosity). (From Bennett, R.H., 1976, Clay fabric and geotechnical properties of selected submarine sediment cores from the Mississippi Delta: Ph.D. diss., Texas A & M University, 269 p.)

FIGURE 3.6 Atterberg testing device, for the determination of the liquid limit of a sediment sample. (Courtesy of SOILTEST, Inc., Evanston, Ill.)

both the intergrain friction and cohesion of a soil sample. For instance, a clay only has cohesion that is an internal physiochemical attraction of the very small (less than 2 μm) clay particles.

Vane shear tests are performed manually in the field (fig. 3.7) or by a motorized device in the laboratory, by inserting a small 90° vane blade of known area into an undisturbed sediment sample and applying a measurable torque to the sample until failure occurs. The amount of torque applied at failure is proportional to the shear strength of the material. If the sample is remolded, a remolded shear strength value can then be measured. The ratio between the two (undisturbed and remolded) provides a measure of the sensitivity of the sediment. This measure becomes an important factor when assessing the strength of a marine sediment following disturbance, such as an earthquake, mud slide, or under cyclic loads (such as imposed by wave action).

The shear strength may also be obtained by the miniature cone penetrometer device whereby the amount of penetration by a small (miniature) cone of known size (angle) and weight, dropped into a sediment sample, is related to the sample's shear strength.

Larger versions of these devices may be used in the field where they may be applied in-place for a more reliable assessment of the sediment strength compared to the testing of disturbed samples brought to the laboratory.

FIGURE 3.7 Hand-held miniature vane shear testing device, for obtaining sediment shear strengths in the field. (Courtesy of SOILTEST, Inc., Evanston, Ill.)

Direct Shear Test

A more complex method of assessing the shear strength of a sample in the laboratory is the *direct shear test* whereby a shearing force is applied to a portion of a sample confined by a split ring or box. The test allows the imposition of a load or vertical stress on the sample to duplicate the load to which the sample may have been subjected in the field. For example, if a core sample of marine sediment were retrieved by a small rotary rig from a depth of 100 m below the sea floor, the equivalent overburden (100 m) could be duplicated by the shear box device. Such vertical stresses may also be applied during tests performed on consolidometer and triaxial devices.

Consolidation Tests

Consolidation tests allow the computation of the amount of settlement that may be expected to take place for a soil sample under given stress or load conditions. A small sample is placed within a confining ring (fig. 3.8) and subjected to ever greater loads on the vertical axis, while the amount of settling, with respect to time, is measured. A plot of the results is referred

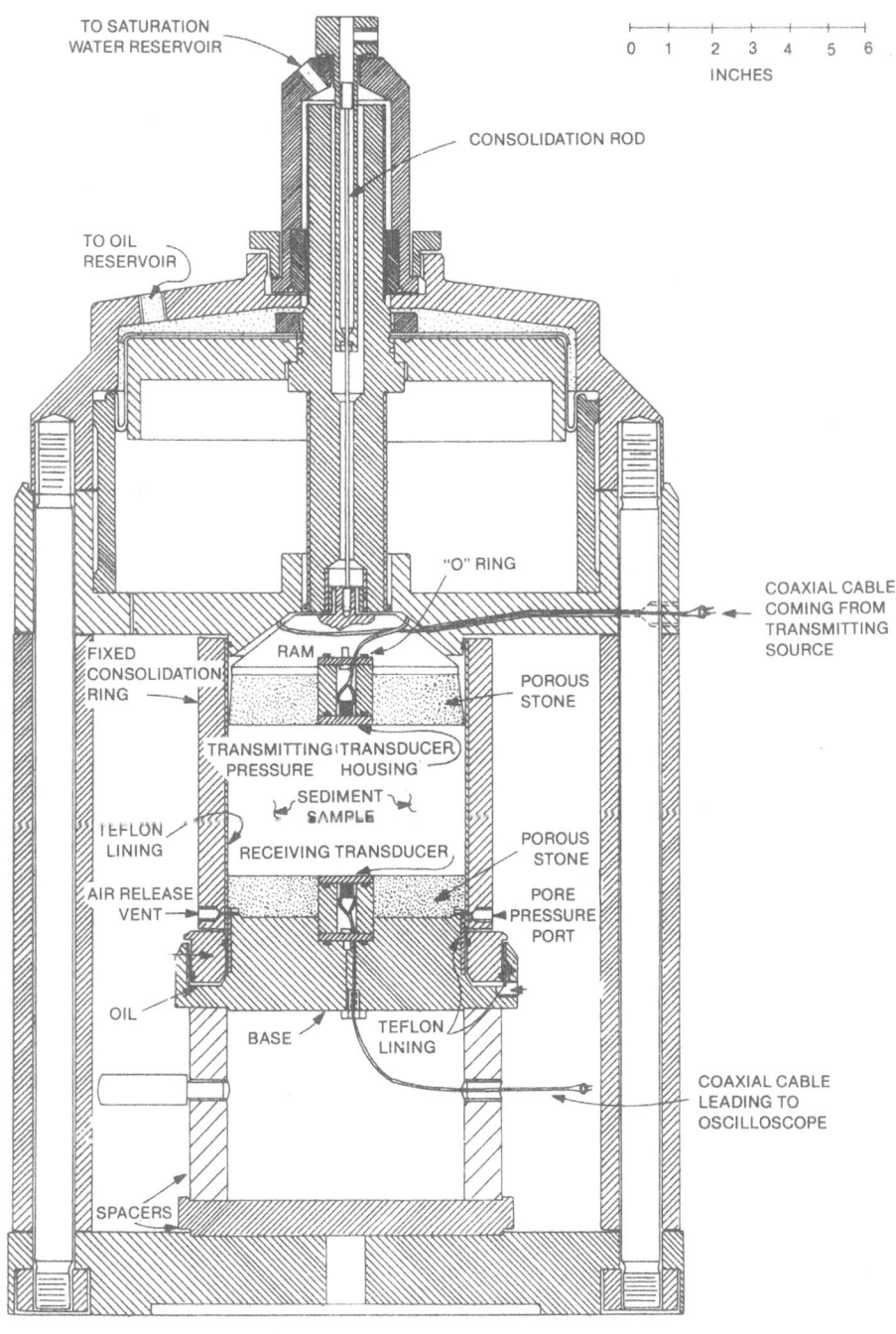

FIGURE 3.8 Diagram of laboratory consolidometer (odometer) apparatus, with transducers for obtaining acoustic velocities. (From Cernock, P. J., 1970, unpublished Ph.D. diss., Texas A & M University.)

38

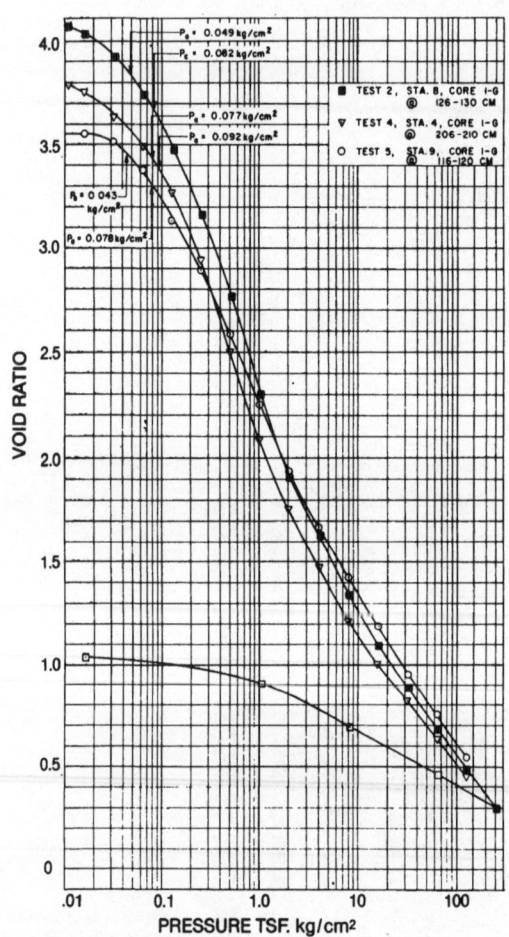

FIGURE 3.9 Curve of void ratio (e) versus log of pressure (P), for typical marine clay from Gulf of Mexico. (From Trabant, P. K., 1972, Consolidation properties of marine sediments retrieved by the *Glomar Challenger*, from the Gulf of Mexico: unpublished M.S. thesis, Texas A & M University.)

to as an e-log P curve, as illustrated in fig. 3.9. This curve may be employed by the civil engineer to compute the amount of settling that would be expected for a given load. Thus, if the stress applied to the soil by a jack-up rig were XXX kPa, the soil would yield by expelling a percentage of its water, and decreasing its void ratio (consolidation) over a given period of time, allowing for settlement of the rig. The permeability or ability of a sediment sample to allow the passage of fluids may also be determined from the results of the consolidation test.

Acoustic Velocities
A specially equipped consolidometer (fig. 3.8) allows the measurement of *acoustic velocities* through a sediment sample as it is subjected to increasing loads, equivalent to greater depths of burial. Such tests have provided insight to the acoustic nature of marine sediments.

Triaxial Compression Testing
For detailed tests of marine sediment samples, where critical loading effects must be assessed, samples are tested within the *triaxial compression testing* device, which allows for very fine control of the environmental stresses applied to the sample up to failure. The sample may also be subjected to cyclic loading during the triaxial test to duplicate the effect of waves or earthquakes passing over the sea floor.

IN-PLACE TESTING OF SEDIMENTS
The in-place or in-situ testing of sediments has gained popularity over the past decades because it represents a more realistic measure of soil properties, since one avoids disturbing the sediments in the sampling process (i.e., coring) or during transport to the laboratory. Large vanes and cone penetrometers with appropriate mechanical remote measuring instrumentation are used during the drilling of engineering boreholes. Results to date have demonstrated their usefulness toward determining the geotechnical properties of marine sediments.

PRESSURIZED SAMPLES
Along a similar line, efforts have recently been made to obtain pressurized core samples from beneath the ocean floor (12) to reduce the effects of disturbance, and capture gas samples by maintaining the in-place hydrostatic pressure on samples for laboratory testing. Samples obtained by this technique require that laboratory testing take place within the confines of a hyperbaric diving chamber, maintained at the original in-place hydrostatic pressure of the sample being tested.

FIGURE 3.10 Free-fall marine penetrometer (MPS-1), prior to launch. (Courtesy of Sandia National Laboratories, Albuquerque, N.M.)

CLASSIFICATION OF MARINE SEDIMENTS

Sediments may also be classified by a number of methods according to the values of their physical properties. The classification scheme presented in Table 3.1 uses several of the physical properties discussed above to place a particular sample into a well-defined category according to the Unified Soil Classification System. Such classification gives the engineer a useful, yet simple, description of a particular soil or marine sediment. For example, a two-letter description of a highly plastic marine clay (non-calcareous) as being of category "CH" would indicate the parameters and properties as listed in table 3.1.

FUTURE DEVELOPMENTS

Present-day research and development programs are being conducted toward the development of marine penetrometers as shown in figure 3.10 (13). Such free fall or propelled devices transmit the response of an accelerometer to a surface vessel as they penetrate the sea floor to depths of 100 m. The recorded deceleration trace permits the assessment of certain physical properties beneath the sea floor without the drilling of more costly boreholes. Besides the measurement of the deceleration, other properties including acoustic velocity, pore pressure, and gamma ray logs may be obtained during penetration.

DESIGN APPLICATIONS

As stated above, the geotechnical properties provide input toward the calculation of the amount of settlement and bearing capacity of a particular portion of the sea floor for an anticipated load, such as may be imposed by a gravity structure or by piles (large diameter steel pipes). While such calculations are the realm of the civil engineer, they are based upon the results of theoretical computations and past field experience.

Using the insertion of a steel pile as an example, the depth of insertion into the sea floor is a function of the friction of the pile surface area with the surrounding sediments, and the bearing capacity (resistance) of the area at the tip of the pile. In turn, these elements are calculated on the basis of the area of the pile (diameter and length) and the shear strength of the sediments. Given the designed or expected horizontal (waves, winds, and sea-floor movements) and vertical loads (weight of structure), to which the pile would be subjected during its lifetime, calculations, including a safety factor, provide the minimum depths to which the pile must be inserted.

GEOENGINEERING SURVEYS

It is thus very important that design engineers be aware of any local variations of the geotechnical properties that may adversely affect their calculations, and which should be indicated by the results and interpretation of a comprehensive HRG survey.

Four
ECHO SOUNDERS

INTRODUCTION

The echo-sounding technique, developed by Professor Fassenden in 1914, has become widely used aboard all seagoing vessels, and is an integral part of all geoengineering surveys. The depths at an offshore construction site must be accurately appraised for the successful installation (even temporary) of offshore drilling vessels, exploration and production structures, and pipelines. An error in bathymetry could result in serious construction problems, since many facilities are designed and constructed for a specific water depth. An example would be the improper location (such as below local sea level) of docking facilities, which are welded to a structure prior to their emplacement offshore. Errors in bathymetry would also result in insufficient water depths for the emplacement of a deep draft jack-up platform. An error of one meter water depth in dredging operations may require the removal of many additional meters of fill, at great additional cost and time.

Hydrography is the science dealing with the determination of water depths toward the production of nautical charts. The end product of a hydrographic survey is a chart of the surveyed area, with bathymetric (depth) contours (isobaths) indicating the water depths, relative to a known datum elevation. Such charts also indicate the slope of the sea floor, the location of shallows, deeps, reefs, and other morphological features that would affect anticipated construction operations.

The echo-sounding unit illustrated in figure 4.1 is typical of the basic systems employed for more than 50 years toward the acquisition of continuous water depth soundings, along the path of a survey vessel. The water depths are generally recorded on a graphed paper strip chart as shown in figure 4.2.

Modern state-of-the-art echo sounders provide digital and paper chart displays of water depth, and permit recording of data in digital format on magnetic tape. It should be noted that all echo-sounding equipment is subject to certain errors and operational limitations. The operator must have a clear understanding of the characteristics and limitations of the instrument.

The following paragraphs outline such factors and discuss corrective measures. The manufacturer's manual should be consulted for details on operation, checks, and maintenance procedures of specific equipment.

TRANSMISSION OF SOUND IN WATER

Water depths are obtained by the echo-sounding technique, by measuring the elapsed time between the transmission of a short acoustic pulse and

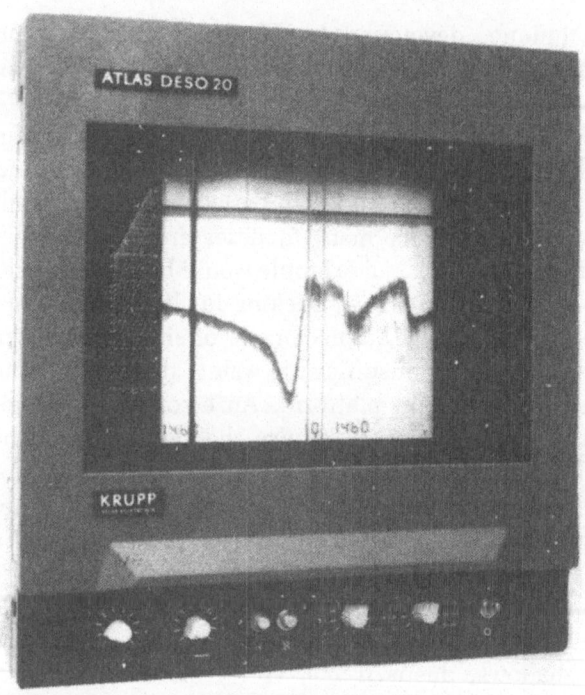

FIGURE 4.1 Photograph of modern echo sounder recording system. (Courtesy of Krupp-Atlas Co., Rahway, N. J.)

the return of a reflection or echo from the sea floor (fig. 4.3). To implement this simple operation a number of problems must be overcome. Among these are:

1. *Noise*, or unwanted sounds that mask the clarity of the returned echo. Noises are generated in the marine environment by organisms (fish, crustaceans, and cetaceans), by ship propellers (cavitation), breaking seas, movement of the survey vessel's hull through the water, and even other acoustic systems.
2. Particles in the water column also generate unwanted noise by reflecting acoustic signals (fig. 4.3), also referred to as the back-scatter effect.
3. The transmitted acoustic signals are further attenuated by their physical *spreading* in a spherical pattern (as waves emanate from a disturbed spot on a pond) away from the transducer and reflecting points in the water column and along the sea floor.

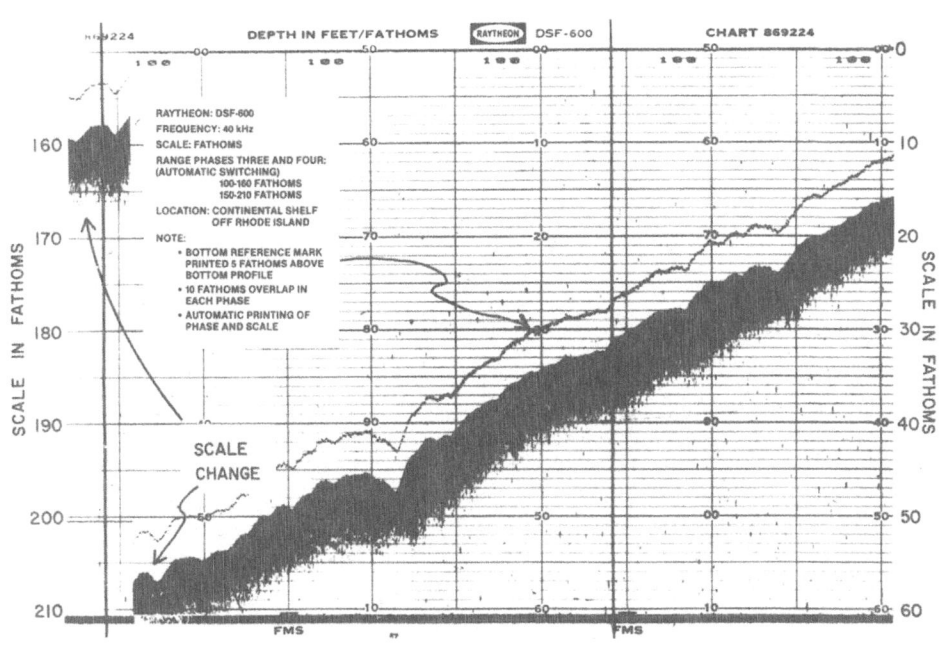

FIGURE 4.2 Typical echo sounder recording. (Courtesy of Raytheon Co., Portsmouth, R. I.)

These problems are overcome by increasing the signal-to-noise ratio (the less noise with respect to the recorded signal, the higher the ratio). This enhancement is obtained by such techniques as:

1. The utilization of discrete frequencies at signal transmission, and by filtering the returned echoes to pass only that particular frequency before amplifying the signal.
2. Amplification of the returned signal, in proportion to its loss by spherical spreading.
3. Reduction of the transmitted beam width (a function of frequency, i.e., the higher the frequency the narrower the beam width), which helps to enhance the signal-to-noise ratio by concentrating the acoustic signal within a narrower cone of transmission (transducer beam width).

SPEED OF SOUND

The speed of sound, which varies as a function of depth (pressure), temperature, and salinity of the water column, must be accounted for to

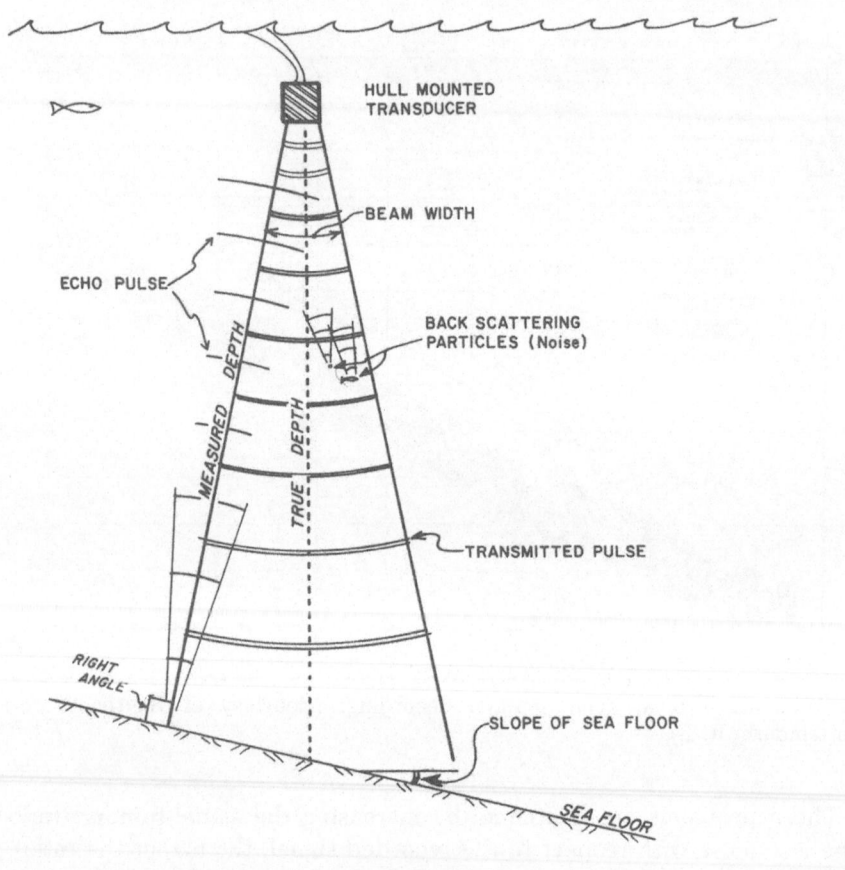

FIGURE 4.3 Acoustic transmission pattern from echo sounder transducer, and effects of beam width. (Drawing by the author.)

obtain accurate survey results. These factors are formulated by:

$$V = \text{Elasticity/Density} = \sqrt{\frac{\gamma}{\rho K}} \, ,$$

where V is the speed of sound (compressional wave velocity), which is independent of frequency, γ is the ratio of specific heats of the fluid, ρ is the density, and K is the compressibility of the fluid. The ratio of specific heats is the result of the conversion of the sound wave energy to kinetic energy (1).

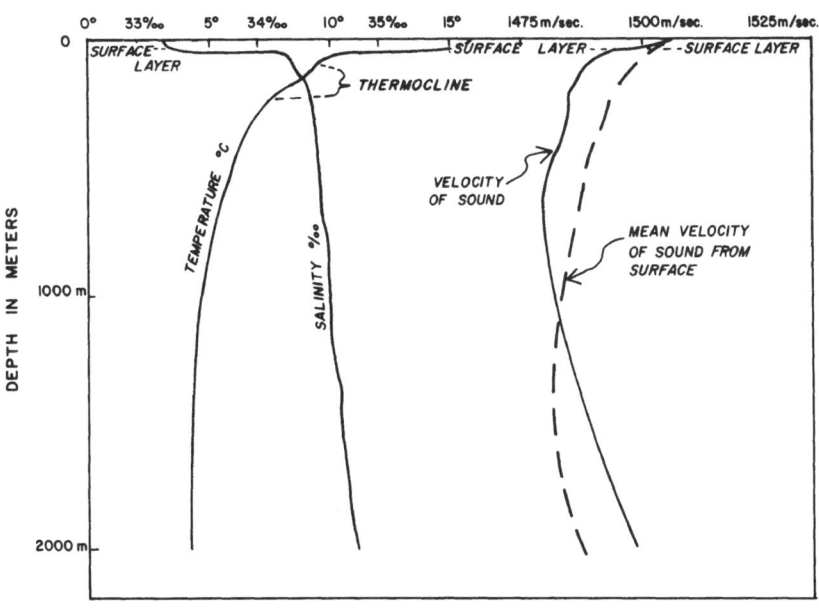

FIGURE 4.4 Typical profiles depicting the variation of acoustic velocity, salinity, and temperature with ocean depth. (Drawing by the author.)

Temperature is found to have the most significant effect at shallow depths above the thermocline, as illustrated in figure 4.4. Pressure is the more important factor at greater depths.

The actual speed of sound in salt water (V) may be computed from the salinity (s), temperature (t), and depth (d) by an empirical equation refined by Del Grosso (2):

$$V = 1410 + (4.21t - 0.037t) + 1.105s + 0.018d \quad \text{(in m/sec)}$$

In practice the velocity employed for computing the water depth should be a mean value for the portion of the water column under consideration, as illustrated by figure 4.4. In shallow water this mean value is that obtained by the bar check calibration procedure, described in a later section. In order to obtain precise water depths from an echo-sounding survey it is imperative that the correct average velocity of sound through the water column be used. It should be noted, however, that the U.S. charting agencies (DMA, NOAA) use 1500 m/sec as their standard (3).

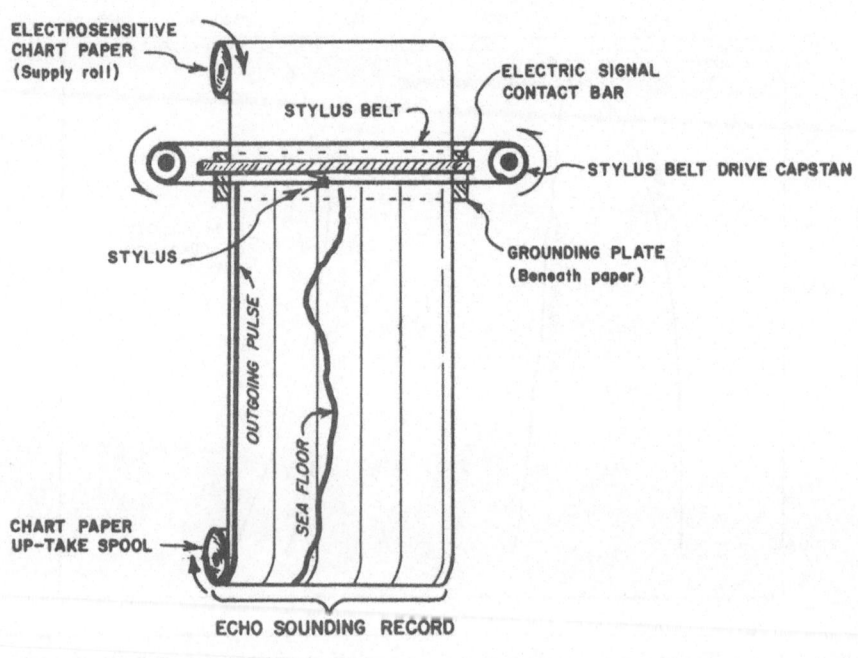

FIGURE 4.5 Block diagram of analog echo sounder recorder. (Drawing by the author.)

PRINCIPLE OF OPERATION (ANALOG SYSTEMS)

Basically an analog echo-sounding system consists of a graphic recorder upon which a stylus-holding belt sweeps across a chart paper (figs. 4.2 and 4.5). As the stylus passes a zero mark (always referenced to time) it triggers an electronic oscillator, which in turn sends a short preselected frequency burst to a transducer (piezo-electric or magnetostrictive) as illustrated by figure 4.6. The recorder marks this event as a thick black line, associated with the transmission. The transducer, thus activated, transmits a brief ultrasonic pulse into the water column at its characteristic oscillating frequency (fig. 4.6). The frequency of operation may vary widely, but generally runs above 10 kHz, depending upon the desired depth range and resolution, as the higher frequencies tend to limit the use to shallower water depths than do lower frequencies.

Returned echoes (reflections) from the sea floor, at the transmitted frequency, activate the transducer (piezo-electric effect) and generate an electronic pulse within the receiver of the echo sounder. The received echo pulse may be very weak, depending upon transmitted power, water

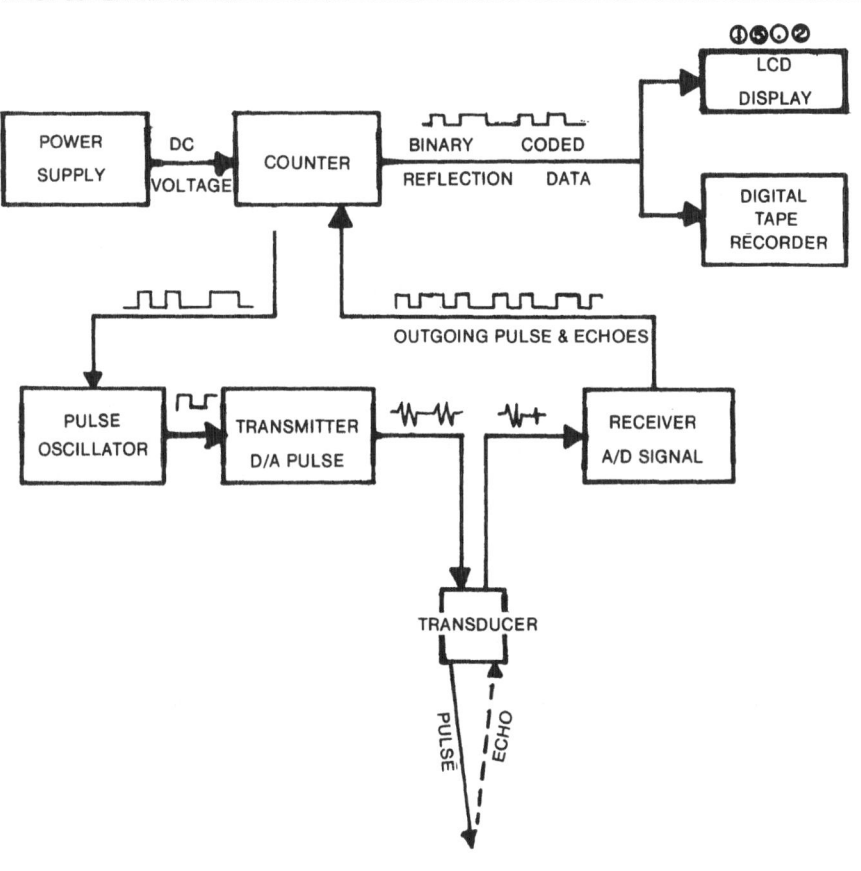

FIGURE 4.6 Digital/analog echo sounder electronic block diagram. (Drawing by the author.)

depth, and the acoustic reflectivity of the sea floor, and must therefore be filtered in order to increase the signal-to-noise ratio, and amplified for display on the recorder by the sweeping stylus, which has moved a distance proportional to the water depth.

The depth is thus produced as a function of the stylus sweep time or rate of travel, by the formula

$$D = VT/2,$$

where D is the depth, V the velocity of sound, and T the recorded travel time (see chap. 2). The two-way elapsed time, due to the signal pulse as it travels down the back, necessitates a division by two.

Figure 4.5 illustrates the basic operational mechanics comprising an analog echo-sounding recorder. The stylus sweep rate determines the scale within analog systems, while digital counters and samplers provide the time base within digital systems.

PRINCIPLE OF OPERATION (DIGITAL SYSTEMS)

Digital systems operate in a similar fashion except that the analog signals amplified by an echo sounder may also be converted to digital binary data, on the basis of a digital time counter, for processing (filtering, gain recovery, and display), and recording on magnetic tape for easier retrieval and manipulations (fig. 4.6).

Thus, digitally sampled data afford rapid display, plotting, or even contouring through computer processing. The use of thermal recorders, which contain no moving parts, is also a common feature of the newer generation of digital echo sounders, as the stylus sweep rate is not used as a reference time base.

RESOLUTION OF ECHO SOUNDERS

The resolution of the recorded echo sounder signals during the course of a bathymetric survey is a function of:

a. pulse length
b. frequency of operation
c. beam width
d. recording technique
e. stylus sweep rate and chart scale

whereby shorter pulse lengths, of the outgoing acoustic signal, increase resolution while limiting transmitted power, and increases in transducer beam width (fig. 4.3) reduce the precision and hence resolution of an echo sounder. As water depths increase, the transmitted beam width and the area of the sea floor covered become larger, reducing the resolution and returned energy levels.

RECORDER PRECISION

The accuracy of the depths recorded is mainly a function of the time base employed to drive the stylus belt or the digital sampling counter. Thus, an

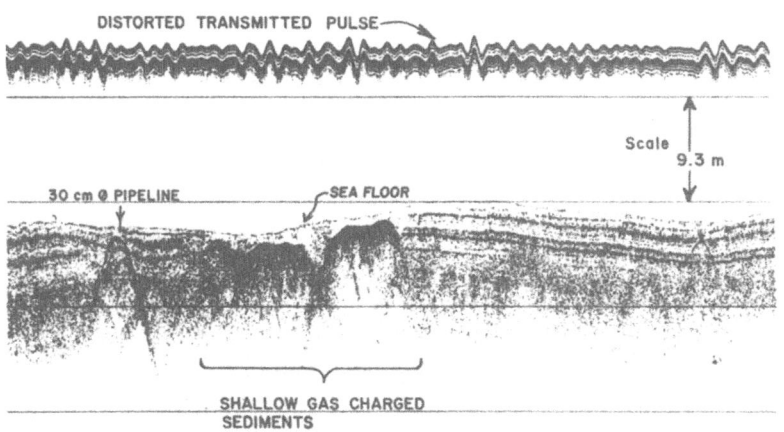

DISTORTED TRANSMITTED PULSE

Scale 9.3 m

30 cm Ø PIPELINE   SEA FLOOR

SHALLOW GAS CHARGED
SEDIMENTS

FIGURE 4.7 14-kHz record depicting effect of heave compensator. Note the distortion of the transmitted pulse as a result of the compensation for the survey vessel motion or "heave." (Courtesy of Ferranti–O.R.E., Falmouth, Mass.)

analog recorder employing 60 Hz alternating current as a time base will produce a 3% error in recorded depths if the time base frequency were to vary by 2 Hz (i.e., 58 Hz instead of 60 Hz). On the other hand, when very high time base frequencies are used, such as in digital- and crystal-based systems, a power supply error of a few Hz is not significant.

MOTION COMPENSATORS

Heave- and motion-compensated echo sounders have been developed for precision bathymetric work where the effects of a survey vessel's motion, such as roll, pitch and heave, would produce poor estimates of the actual water depths. The systems designed to compensate for such effects involve the removal of such motions, by either sampling them with accelerometers, filtering them by sampling over a period of time (swell filters), or stabilizing the transducer on a gyroscopic platform, and applying proper compensatory corrections to the data. Such corrections may be made on either analog or digital data. An illustration of the effects of such a compensatory system is shown in figure 4.7.

The use of gyroscopically controlled motion compensators permits utilization of very narrow beam width transducers, to produce precision soundings of greater resolution and accuracy. Motion-compensating systems are sometimes used aboard larger survey vessels, but not for portable systems aboard small craft.

MULTIPLE BEAM SYSTEMS

Several special systems for obtaining water depths have been developed to provide greater accuracies or larger areal coverage of the sea floor along a single survey line.

A simple form of such a system consists in the deployment of a number of transducers, along a line normal to the survey vessel's course, and the simultaneous recording of the individual channels, as illustrated by the configuration of figure 4.8.

This system is effectively limited by both the distance separating the transducers and motion of the survey vessel. Nevertheless, it provides valuable additional coverage for shallow depths, particularly when integrated with digital recording and processing systems as in the conduct of large-scale survey operations, such as are conducted prior to dredging operations near shore and in harbors.

A multibeam system is a multichannel echo sounder, which, instead of using a single transducer, transmits a large number of pulses along a fan-shaped pattern to either side of the survey vessel (4). These are in turn recorded digitally with respect to their true spatial orientation for computer processing in the data reduction stage. The transmissions are keyed by the input from a gyroscope, in order to maintain an effective beam width within a few degrees of vertical. However, such systems are limited to specially dedicated survey vessels.

The contoured bathymetry of figure 4.9 illustrates the results of such a multibeam survey over a salt dome/reef complex in the Gulf of Mexico, while figure 4.10 depicts the difference in coverage between a single and multibeam survey. The advantage of such systems is their ability to cover a large swath of the sea floor along a survey vessel's track with great accuracy, while reducing the number of survey tracks and cost.

Originally developed for the U.S. Navy in the early 1960s under the acronym SASS (Sonar Array Sounding System), multibeam echo-sounding systems are used mostly for small-scale (large areas) hydrographic surveys to water depths of more than 10,000 m. As petroleum exploration operations reach greater depths beyond the continental shelf, the use of multibeam echo-sounding techniques should become more prevalent in view of their numerous advantages over single beam systems (3).

RECORDER/
TRANSCEIVER
& TRANSDUCERS

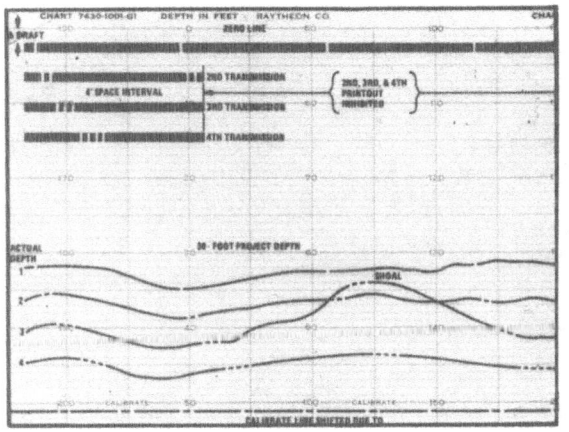

INDIVIDUAL ECHO
SOUNDING TRACES

VESSEL PERFORMING
SWEEP SURVEY

TRANSDUCERS

FIGURE 4.8 Multi-transducer echo-sounding system, recording, and survey vessel. (Courtesy of Raytheon Company, Portsmouth, R. I.)

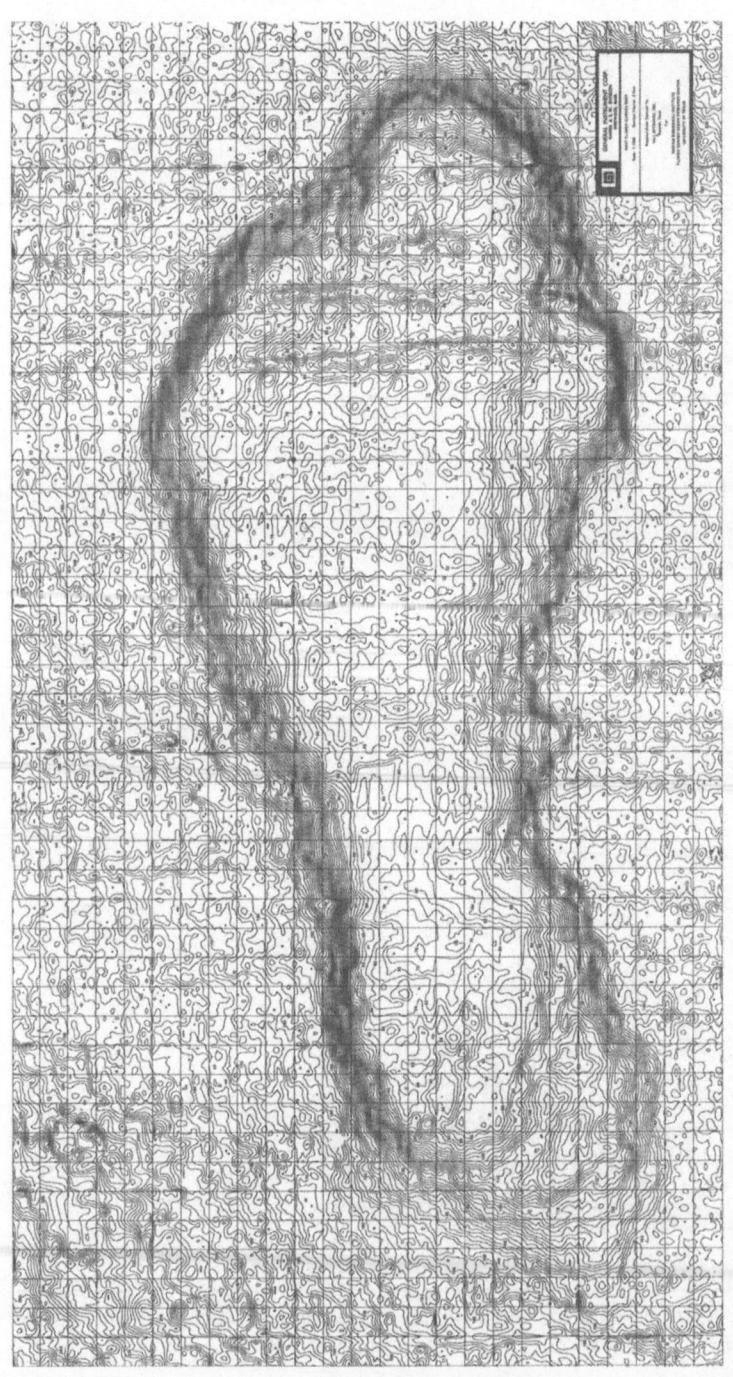

FIGURE 4.9 Bathymetric contours from multibeam survey on Flower Garden Reef, Gulf of Mexico. (Courtesy of Hill Offshore, Inc., Houston, Tex.)

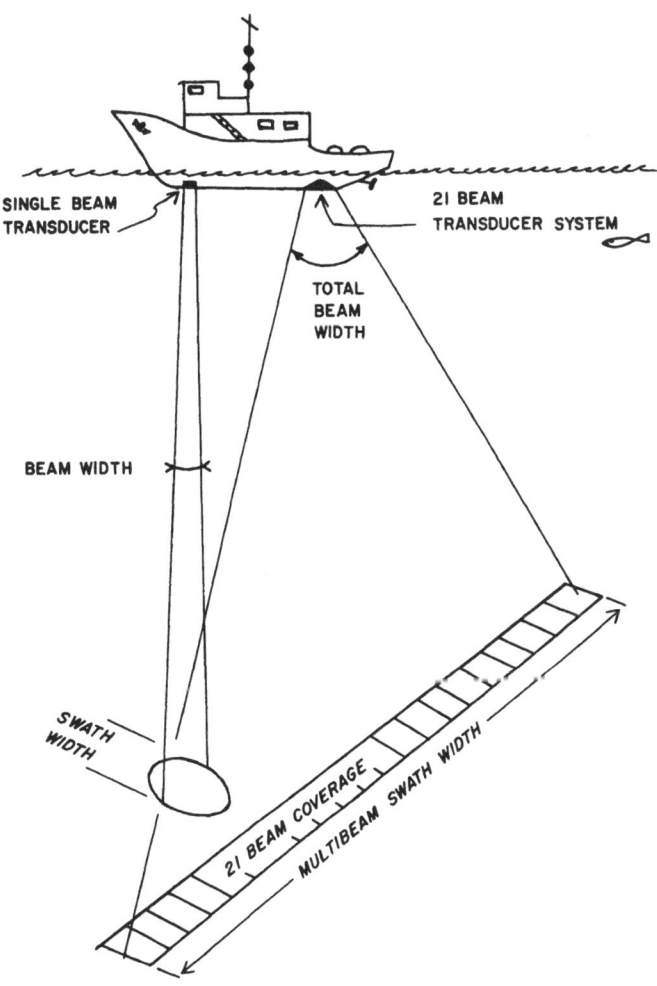

FIGURE 4.10 Comparison of echo sounder survey coverage by conventional single beam transducer, and multibeam "swath" coverage. (Diagram by the author.)

DEEP-TOW SYSTEMS

Towing transducers at depth, with either pressure (hydrostatic) or upward scanning echo sounder transducers, provide more reliable data in great water depths of thousands of meters. This increased accuracy is due to the shorter distance and resultant beam width between the echo sounder transducer and the sea floor.

When deep-tow systems are deployed at the end of several thousand meters of cable behind the survey vessel, the precise location of the

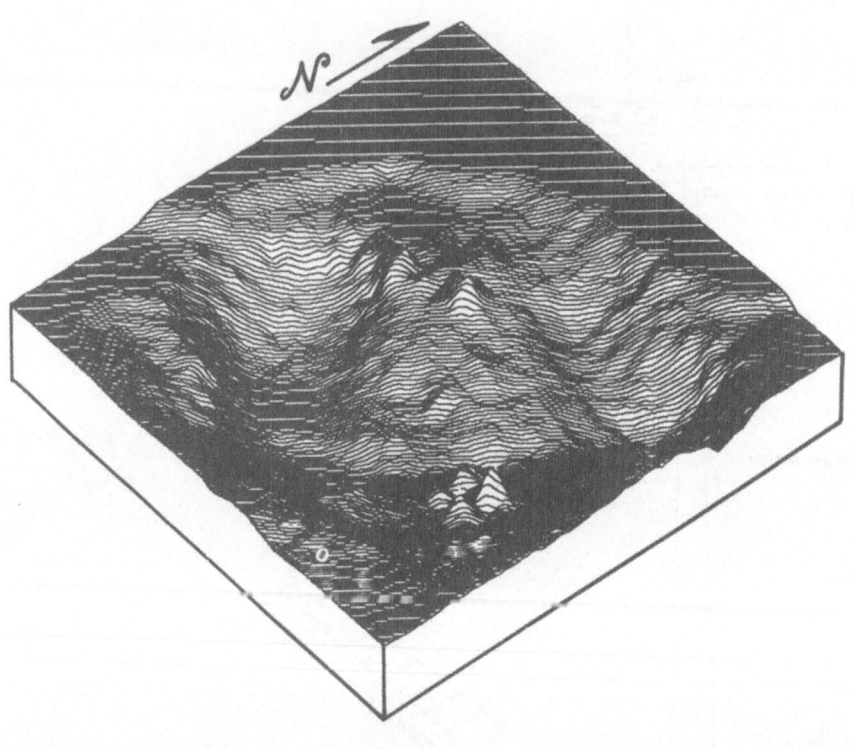

FIGURE 4.11 Example of 3-D, computer processed and plotted, bathymetric display from Orca Basin, Gulf of Mexico. (From Trabant and Presley 1978 [6].)

transducer is needed for accurate results. This factor has imposed severe limitations on the use of the deep-tow technique to date.

The use of deep-towed instrument packages for HRG profiling surveys in deep water has become prevalent over the past decade. The sensors usually include long- or medium-range side scan sonar systems and a subbottom profiler or echo-sounding transducer.

COMPUTER PROCESSING

Digitally recorded echo-sounding data can readily be processed by a computer with appropriate software (fig. 4.9), to produce not only contoured bathymetry, but also three-dimensional displays at oblique angles of one's choosing, as depicted in figure 4.11. However, care should be taken to properly edit the digitally recorded field data for glitches and other typical digital data entry errors, which would produce the proverbial garbage-in-garbage-out effect, and erroneous bathymetric features.

State-of-the-art hardware and software also permit the conversion of analog data records to digital format, through the use of digitizing tables, providing the interpreter with a valuable tool toward the reduction of large volumes of data.

OPERATIONS AND CALIBRATION

Obtaining inaccurate results during the course of a bathymetric survey may be worse than obtaining no data at all. All factors affecting the precision of a bathymetric survey should be checked with care prior to survey operations.

In shallower waters (up to perhaps 100 m), direct calibration of the system is possible. This calibration is performed by lowering an acoustically reflective object such as a steel plate or bar, beneath the transducer on an accurately marked line (nonstretching material), to the approximate depth expected to be covered by the survey at the survey site. When the bar is at the known depth, the recording instrument is adjusted first for the transducer depth (draft) and then the speed of sound control (stylus sweep rate) so that the bar reflection is precisely recorded at its proper depth. It is important that the bar be maintained directly beneath the transducer during calibration.

An alternate method consists in the use of a pressure sensor, employing a large calibrated gauge, and referred to as a pneumatic gauge. By this technique, a pressure sensor is placed on the sea floor beneath the echo sounder transducer, and air pressure applied until equilibrium is reached. The corresponding pressure reading at this time may be converted to depth as a function of the density of the overlying water column. The density, as discussed earlier, which varies with both salinity and temperature can lead to a slight error in the pneumogauge technique. These procedures can provide the correct velocity input for a given survey area. All such field procedures should be annotated in detail on the record section for future postsurvey reference.

Tidal variations at the survey site are best obtained by direct measurement. This usually involves the deployment of a recording tide gauge for the duration of the survey. In order to verify the specific scale in use and eliminate the wrap around problem on an echo sounder recording system, one may briefly change scales to encompass a greater depth.

Annotations on the records or logs should be made for such miscellaneous items as: presence of other vessels in survey area, location of fishing trap and net buoys, surface currents or rip tides, sea state and winds, charted features such as reefs, wrecks, and military ordinance disposal areas, shipping fairways and anchorages, and color changes of the water along the survey tracks. The offset distance between the navigation receiver antenna and the echo sounder transducer should also be noted on the records.

Sounding or survey lines are generally run perpendicularly to the coast line or bathymetric contours, with tie lines at a normal angle (90°) and greater spacing. A general rule of thumb is to maintain the spacing between normal and tie lines at a ratio of 10 to 1. The assembly of the survey lines is referred to as a grid pattern, for which the exact spacing depends upon both the objectives of the survey and the local bathymetry. Thus, a flat sea-floor area would require far fewer survey lines than one with considerable morphology in order to produce an accurate chart. For echo-sounding operations in waters deeper than about 1000 m, the use of deep-towed, narrow, or multibeam systems and migration corrections is a necessity for accurate charting.

INTERPRETATION

A first step in the interpretation of echo soundings should consist in a check of the calibration(s) and recorded depths against existing charts of the area for verification purposes. If side scan sonar data were obtained in conjunction with the bathymetry in shallow water, it is worthwhile to check the echo sounder results against the arrival times of the sea surface and bottom side lobe returns (see chap. 5). Secondly, the repeatable accuracy should be checked at the survey tie-line crossings (intersects) to ensure that all the depths tie in correctly. Errors at tie locations are usually caused by equipment timing changes during the course of a survey, errors in navigation, or incorrect tidal reductions.

If available, the tidal information for the survey area, with respect to time of survey operations, may be inserted on the records. Tidal correction values inserted at the beginning and end of a survey line record are sufficient for the data to be corrected by linear interpolation.

Next, the water depth data are transferred by inserting their numerical values at respective shot points or fix marks on the navigational survey plots. Additional values between fixes should be inserted where changes in water depth, such as peaks or troughs, are indicated. The data should also be adjusted for the sensor (transducer) set-back or offset distance with respect to the navigation antenna position aboard the survey vessel.

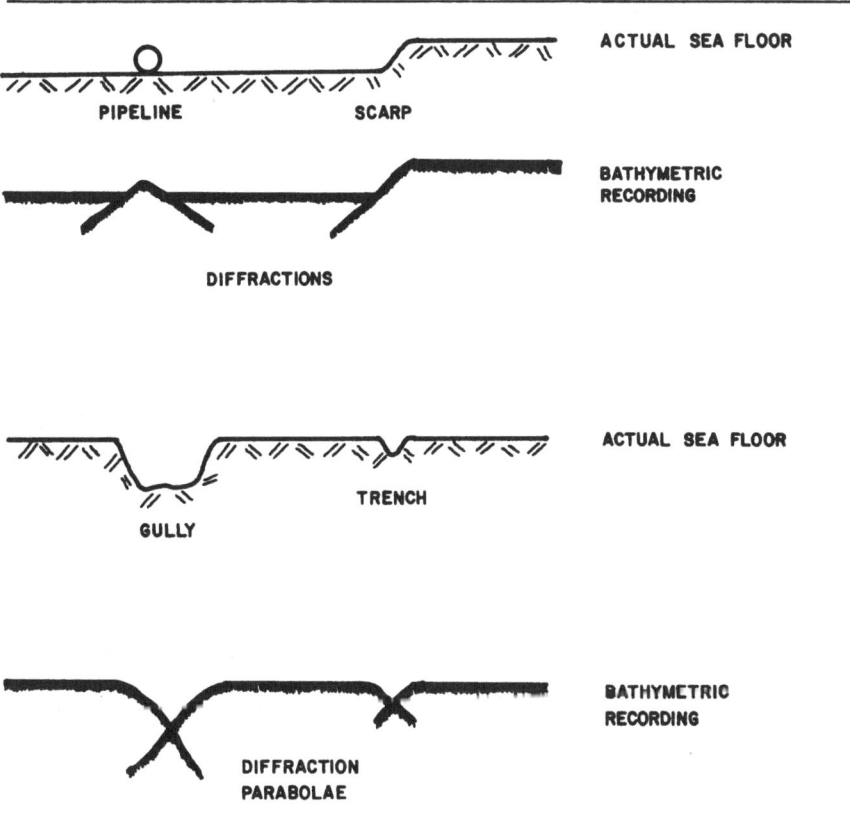

FIGURE 4.12 Effect of angular features on sea floor on echo sounder recordings. (Drawing by the author.)

In the interpretation of echo soundings, one should bear in mind the vertical exaggeration of the records (ratio of horizontal to vertical scale), which exaggerates slopes of the sea floor, and that the returned echoes are from the closest distance to the bottom within the generated acoustic beam cone (fig. 4.3). This latter effect produces slightly shallower depths, where sea-floor slopes attain angles greater than about fifteen degrees (depending upon beam width and water depth). The procedure for removing this effect is referred to as *migration* in geophysics (5).

Similarly, the recorded echo soundings from angular features on the sea floor tend to be smoothed out, as illustrated in figure 4.12.

One should also be aware of the generation of false bottom echoes from acoustic reflectors in the water column such as the Deep Scattering Layer

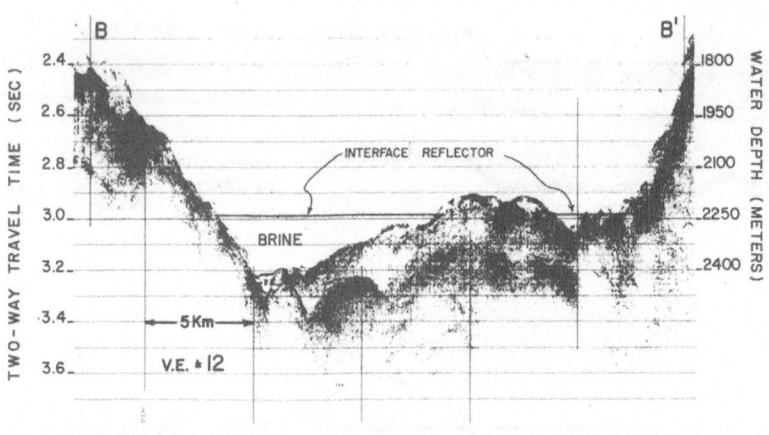

FIGURE 4.13 Acoustic reflector from deep ocean brine pool, Orca Basin, Gulf of Mexico. (From Trabant and Presley 1978 [6].)

(DSL), or fresh/saltwater interfaces such as occur at river outlets and over brines, as shown in figure 4.13.

Care should be taken during acquisition and when interpreting water depths to account for the occasional *wrap around effect,* whereby reflections from depths greater than those being recorded may produce shallower values (e.g., if a recorder is operating on a 40 m scale in a 60 m water depth, the wrapped around signal will show up at a depth of only 20 m.).

Bathymetric charts should be annotated as to scale, contour interval (C.I.) values, and the reference sea level datum employed, generally the local nautical chart datum. Examples of datums are: local sea level at time of survey (relative), Lowest Astronomical Tide (LAT), used on most British Admiralty charts, Lowest Water Level (LWL), Mean Low Water (MLW), Indian Spring Low Water (Arabian Gulf), and Lower Low Water (LLW in the Gulf of Mexico).

Features revealed by echo sounders of concern to the emplacement of structures may include: displacements of the sea floor associated with active faults, presence of mud volcanoes and pockmarks (small depressions), gas seeps in the water column, surface mud flow channels, glacial deposits, iceberg scour, sand waves, and variations of acoustic reflectivity. The geological history of a particular survey area is also helpful in the interpretation, as is the composition of the sediments on the sea floor.

Five
SIDE SCAN SONAR

INTRODUCTION

The side scan sonar represents the only high-resolution seismic tool that provides coverage to both sides of a survey vessel's track. It is a most practical and valuable tool for obtaining an acoustic picture of the sea floor. Operational side scan sonar systems were developed at the British National Institute of Oceanography following World War II, as an application of antisubmarine warfare techniques used toward oceanographic exploration.

The first commercial system, known as the "fisherman's ASDIC," was produced by the Kelvin-Hughes Company. The first reported use of side scan sonar toward geological investigations was reported by Chesterman et al. (1). There are a large number of systems in use today for offshore geoengineering surveys. While these systems have gone by a variety of names such as ASDIC, BASDIC, and sideways looking sonar, the term side scan sonar has received wide acceptance during the past twenty years of utilization. Sonar is an acronym for SOnic NAvigation and Ranging. Side scan sonar records are commonly referred to as sonographs (2).

APPLICATIONS

Side scan sonar units permit the detection of bathymetric irregularities and objects on the sea floor. Interpretation of sonographs allow complete mapping of such features as geologic outcrops, variations of surface lithology, sunken wrecks, and other items of concern to the emplacement or construction of structures on the sea floor. Useful applications are found daily in the course of offshore development, in particular its use as a search tool for items lost on the sea floor. Ecologists and marine biologists may also employ sonographs toward faunal assessment of the sea floor and water column, by detecting fish schools, kelp beds, and oyster reefs.

PRINCIPLES OF OPERATION

The principle of operation consists in the transmission of high-frequency (10 to 500 kHz) sound pulses along narrow fan-shaped beams slightly depressed (10° to 20°) with respect to horizontal (sea level), from the transducers pointed to either side of a survey vessel's track (fig. 5.1). The recorded echoes produce an oblique plan view of the morphology and texture of the sea floor as a function of acoustic reflectivity, within the limits of a system resolution. Typical beam widths are on the order of a few degrees, and pulse lengths are about one msec in length for operation

FIGURE 5.1 Towed side scan sonar and subbottom profiler in operation, with acoustic transmission beams depicted. (Courtesy of Klein Associates, Inc., Salem, N. H.)

on a range of 1000 m (table 5.1). The beam angle within the vertical plane is not overly critical as weaker secondary beams (side lobes), which reach the sea floor and sea surface at angles of up to 90°, tend to give adequate coverage to nearby features within these zones (fig. 5.1). A narrow horizontal beam is, however, a critical requirement for resolution and long-range transmission, since it helps concentrate the transmitted acoustic power.

The transmitted sound pulses and returned echoes travel at the speed of sound in the water, just as with echo-sounding equipment (chaps. 2, 4). The physics involved in sonar transmission may be found in detailed texts on the subject (3).

A detailed compendium on side scan sonar operation and the interpretation of sonographs has been prepared by B. W. Flemming (4), Flemming et al. (5), and a practical descriptive atlas of recorded sonographs has been assembled by Belderson et al. (2).

RESOLUTION

The resolution obtained by side scan sonar systems is a function of the signal beam width, frequency, pulse length, and recording method. In practice, resolution is equal to approximately one-thousandth of the operating range (4). Thus, at a frequency of 100 kHz the computed resolution is on the order of 15 cm on a range of 300 m, and 7 m for a 6.5 kHz system on a recorded range of 22 km. Additional factors affecting resolution include pulse repetition rate, ship speed, and the dynamic range of the paper recording medium.

One must also consider the resolution of the recordings, which are dependent upon sweep rate density, and the dynamic response of the recording media. Both transverse and horizontal resolution must also be taken into account. Transverse resolution is equal to the minimum distance between two objects parallel to the line of travel that will be recorded on paper as separate objects. Vertical resolution is the minimum distance between two objects perpendicular to the line of travel that will be recorded on paper as separate objects.

Although these factors may provide theoretical limitations in resolution, the acoustic reflectivity and target coherence are frequently more important determining factors. Thus a very small object, such as a small diameter cable, may still produce a strong reflector, which may not have been anticipated on the basis of theoretical computation, because it is a long coherent target.

TABLE 5.1 Major side scan sonar systems and parameters

Category/ Manufacturer	Model	Frequency (kHz)	Max. Rge. (meters)	Pulse Length (msec.)	Main Beam Angle	(H/V degrees)
LONG RANGE N.I.O.*	GLORIA	6.5	22,000	12	2.7	10
MID RANGE CGG	SOL-120	23	2,400	2	2.4	25
Ferranti O.R.E.	163	30	2,000	1.2	2.5/2.0	55
I.S.T.	Sea-Mark I	2,500	var	1.7		
I.S.T.	Sea-Mark II	5,000	var	2.0		
SHORT RANGE E.G. & G.	Mark 1B	105	500	0.1	1.2	20 ?
Klein	520	50, 100 & 500	600	0.1	1.0	40/20
Ferranti O.R.E.	1500	100	500	0.2	1.1	18/28
EDO Western	606	100	400	NA	NA	NA
WESMAR	500SS	105	480	0.1	1.5	35
Kelvin Hughes	MS 43	48	v 550	1.0	1.5	51

*National Institute of Oceanography, Great Britain.

An additional consideration in the interpretation of medium- and long-range side scan sonar data is the penetration of the uppermost sediments, by the relatively low operating frequencies. This penetration tends to produce a sonograph of the underlying strata instead of the actual sea floor, where it is composed of loose fine-grained sediments.

In practice, just as with bottom-penetrating seismic systems, no single device will meet the conflicting requirements of range and resolution, so that it may be necessary to use more than one side scan sonar system to obtain both wide coverage and detailed resolution.

SIDE SCAN SONAR SYSTEMS

Side scan sonar systems are generally composed of three components: a towed transducer fish and cable, a dual channel paper recorder, and transceiver electronics package (usually enclosed within the recorder) as shown in figure 5.2. For systems operating in a digital mode, an additional electronics package and magnetic tape drive/recorder may also be included.

The sideways pointed transducers are normally housed within a hydrodynamically streamlined fish, which is towed behind, to the side, or for some applications (shallow water) ahead of a survey vessel. The reflected echoes are amplified and recorded graphically on the dual channel recorder (fig. 5.2). The origin (time zero of pulse transmission) is located in the center of the recording, as the sweep moves simultaneously outward to the maximum range of both sides (fig. 5.3 [top]). As the amplified electrical current passes from the printer blade electrode through the recording paper to ground, marks of varying intensity are made in proportion to the strength of the incoming signal (fig. 5.3 [top]).

The returned echoes are usually preamplified within the transducer fish, and again at the recording instrument. In addition to linear amplification, the incoming signals are increased in gain with respect to time in order to compensate for signal loss due to spherical spreading through the water column and other factors (referred to as TVG, or time varying gain). The amplified signals are then printed on the graphic recorder as a continuous recording of acoustic features along the survey vessel's track (fig. 5.3 [top]).

On the basis of operational range and frequency, one may place side scan sonar systems into one of the following three categories (table 5.1):

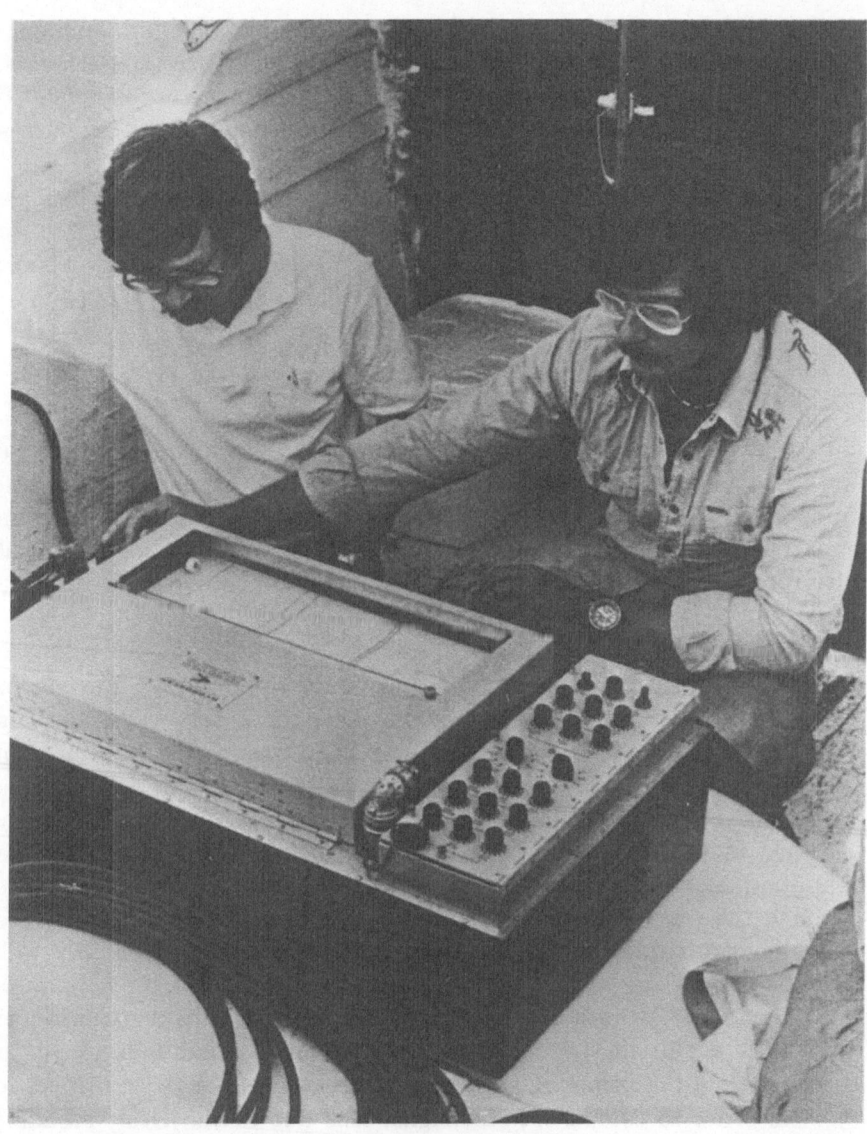

FIGURE 5.2 Three-channel side scan sonar/subbottom profiler recorder and transceiver system during shipboard survey operation. (Courtesy of Klein Associates, Inc., Salem, N. H.)

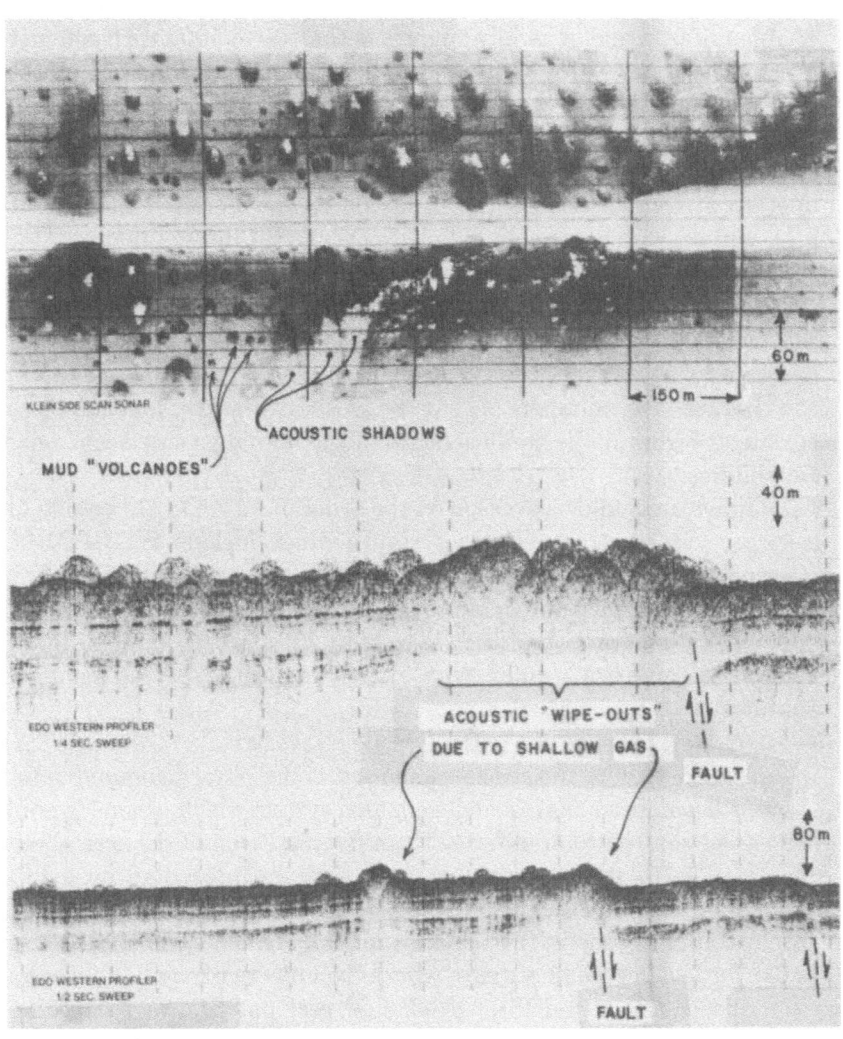

FIGURE 5.3 Sonograph (*top*) and subbottom profiler record (*bottom*) showing large field of mud "volcanoes," northern Gulf of Mexico. (Courtesy of Summit Geophysical, Inc.)

1. *Short-range* systems provide coverage of the sea floor out to distances of about 300 m, and generally operate in the 100 to 500 kHz frequency range. These units are the most commonly employed for work on the continental shelf. A few miniature units are sufficiently portable to allow operations in shallow bays and rivers from small boats. Also included in this category are the vary high resolution systems that operate at frequencies on the order of several hundred kHz (table 5.1), and provide extremely detailed coverage of the sea floor.
2. *Midrange* side scan sonars have ranges on the order of a few km, and operate at frequencies between 10 and 80 kHz (table 5.1). Systems of this type are becoming very useful for survey work beyond the shelf break, where water depths exceed 200 m, and wider range coverage can reduce the number of survey lines necessary for complete coverage. Figure 5.4 is an illustration of the midrange side scan sonar capabilities.
3. *Long-range* units allow coverage on the order of 20 km to either side of a survey vessel's track. These systems, which include the GLORIA (Geological LOng Range Inclined Asdic [2]) are used primarily as an oceanographic tool toward marine geological investigations. The units operate at frequencies under 10 kHz, with a resolution on the order of kilometers, and are usually very bulky, requiring special handling equipment.

An additional tool, to the above categories, is the *sector scanning sonar*, or obstacle avoidance sonar, a single channel system which, when lowered from a vessel, may be rotated for a circumferential scan of the area about the vessel. The rotational scanning sonar data are presented on a PPI (Plan Position Indicator) scope operated in a manner similar to radar, and have found applications in the fisheries industry. Such units should also be of use for shallow water surveys where the objective may be to clear a passage ahead of a vessel through shallow reef patches, or to observe ongoing activity beneath the sea surface, as in the emplacement of cables and pipelines or production platform jackets. Side scan sonar systems may also complement *deep-tow packages*, or be installed within a tow fish containing a subbottom profiler, as illustrated by the recording of figure 5.5.

DEEP-TOW SYSTEMS

Deep-tow side scan sonar units of the midrange type are being used for survey work in the deeper waters beyond the continental shelf edge. The

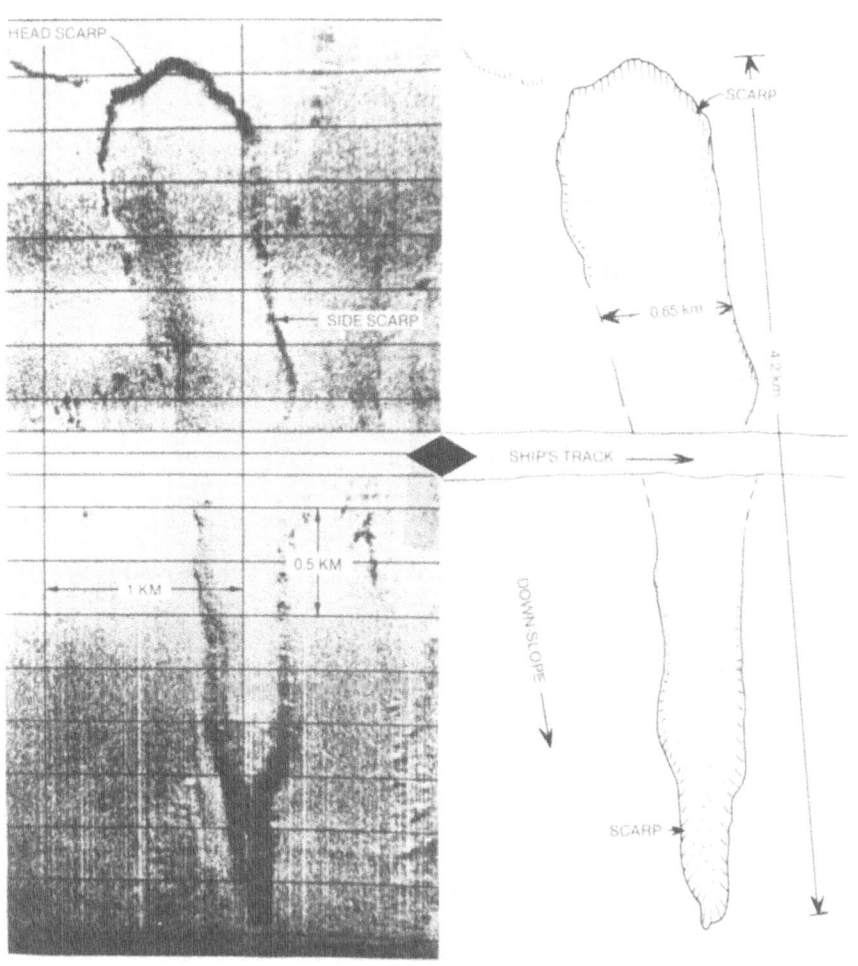

FIGURE 5.4 Sonograph and interpreted drawing (*right*), of elongate erosional feature, at base of continental slope off Georges Bank, northwest Atlantic, utilizing a midrange side scan sonar system (Sea MARC I). (Open file data, courtesy of Office of Marine Geology of the Mineral Management Service, U.S. Dept. of the Interior, at Woods Hole, Mass.)

tow fish package is deployed on long cable lengths, which may exceed lengths of 3000 m for operation in water depths of one to three km.

The major drawback in the use of deep-tow systems is the lack of precise location of the transducer fish and resultant recorded data. While tow depressors are used to shorten the necessary cable length deployed

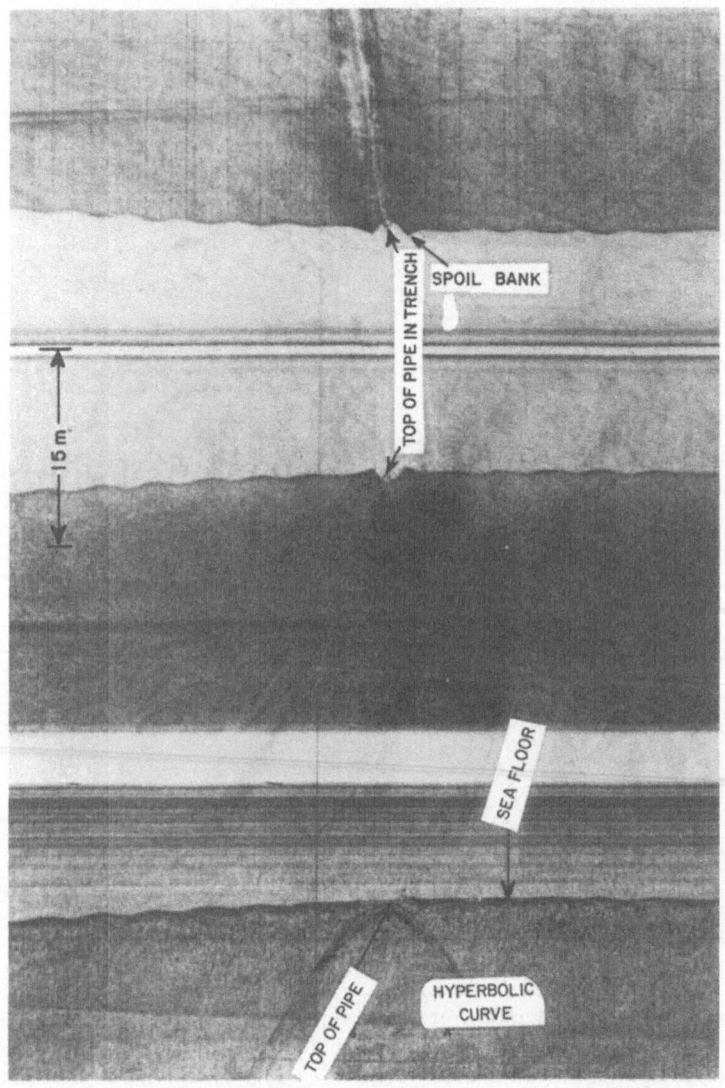

FIGURE 5.5 Three-channel combination side scan sonar (*top*), and subbottom profiler record (*bottom*), showing pipeline in trench, from North Sea, O.H.P., Ltd. (Courtesy of Klein Associates, Inc., Salem, N. H.)

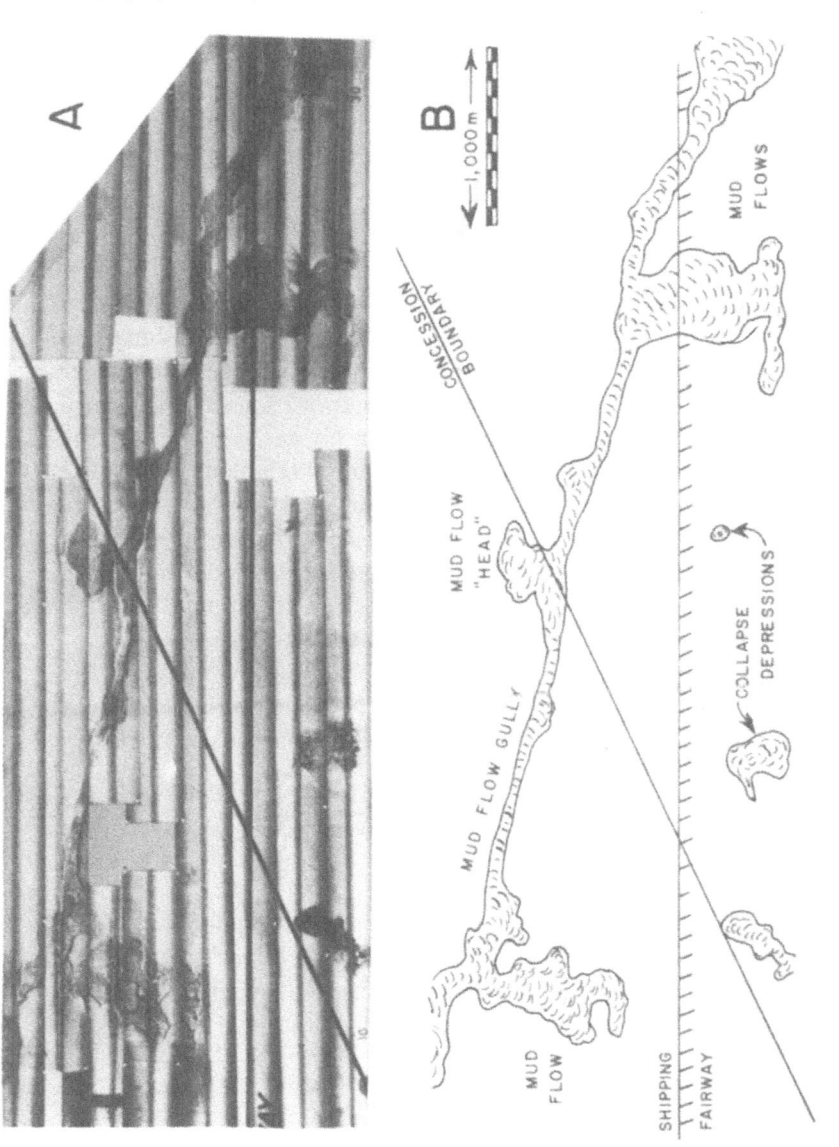

FIGURE 5.6 *A.* Photo-mosaic assembled from slant-range corrected sonographs. *B.* Interpretation depicting collapse depressions and mud flow system, Mississippi River Delta (EG&G, SMS 960). (Courtesy of Hunt Oil Company, Dallas, Tex.)

(6), only acoustic navigation or positioning systems appear to offer a solution to the problem. The principal system used for the positioning of deep-tow packages involves obtaining range and bearing information relative to the survey vessel by acoustic transponders.

DIGITAL OPERATION

While most of the side scan sonar systems were originally designed for analog operation, they may all be modified to digitize their signals to facilitate image recording. Digitized data allow the removal of the water column (height of fish above sea floor), and application of slant range corrections. Such techniques permit the assembly in real time of true planimetric side scan images, which may be assembled into sonar mosaics, having equal scales along and to the sides of the ship's track, to depict large areas of the sea floor as shown by the photo-mosaic in figure 5.6. The necessary software for slant range corrections has been developed by a number of companies, and is readily applied through digital storage in memory within a microprocessor.

Recorded data of this nature may also be played back within an array processor computer, such as employed in the reduction of multifold seismic recordings, for the production of side scan sonar mosaics covering large areas of the sea floor.

OPERATIONAL PROCEDURES

Details as to operational procedures for specific equipment, from installation to fine tuning, may be found in the manufacturer's manual and should always be consulted. The following items are of general concern in the deployment and operation of side scan sonar systems.

Tow Methods

The side scan sonar transducer housing or fish is usually towed astern, by a conductor tow cable. The amount of cable deployed is a function of the survey area water depth, and vessel speed. The optimum height of the tow fish package off the sea floor is typically on the order of 10% to 20% of the maximum lateral coverage (range). In very shallow water survey areas, where the vessel's propeller and noise may interfere with deployment and operation, the unit may be towed from the bow of the vessel.

Turns, at the end of survey lines, must be made with care as the tow fish units tend to sink. Abrupt changes in sea-floor elevations must be anticipated when determining the amount of tow cable. Adequate winching facilities may be necessary to pull the unit in (up) when elevation changes warrant. The use of slip rings, which allow electrical contact between the cable and recording unit while the winch is operated, are necessary to allow continuous data recording as the tow fish is let out or in. The amount (length) of cable deployed is generally logged on the recordings in order to assist computation of the lay-back correction of the recorded data with reference to the navigation antenna.

Most side scan sonar transducers are equipped with emergency breakaway tail fins and primary cable couplings. Thus, in case the unit drags on the bottom or upon contact with an object or the sea floor, the fins are released and a tripping mechanism frees the unit, allowing it to be retrieved.

Side scan sonar units which provide corrected, isometric, or true planimetric sonographs in real time, require an input for the speed of the vessel. This may be done by the use of a separately deployed log (speed) sensing impeller), the survey vessel's speed log, or an input from the navigation system. The latter input provides the true ground speed of the vessel, with respect to the sea floor, and is not affected by currents and drift of the survey vessel.

The tuning of side scan sonar recordings is very delicate and requires considerable hands-on experience to produce picture-perfect sonographs. A fine balance between the overall gain, TVG onset, and slope are critical toward this end. TVG compensates for the signal losses associated with the effects of spherical spreading of the acoustic beam. Automatic gain settings, available on certain units, are generally not adequate for the generation of detailed sonographs. Subtle variations of the sea floor, such as occur between sands and clays, may be masked by the automatic gain changes.

Survey grid patterns employed for the use of side scan sonar coverage should be calculated to provide a slight overlap between adjacent survey lines of the side scan sonar's effective maximum range (4).

Such overlap produces 100% coverage of the survey area, while for certain survey conditions requiring very detailed coverage, such as in the search for very small objects, an overlap of more than 100% may be warranted. Thus, looking at features or targets from several angles and sides may be necessary to obtain a complete image of a particular feature.

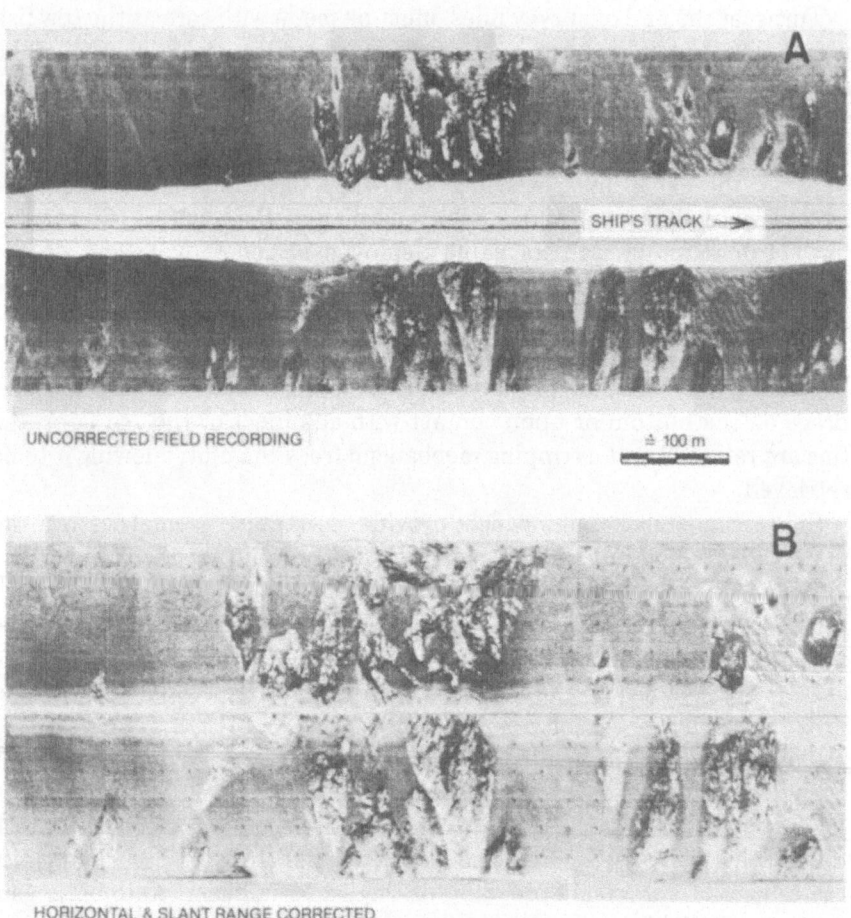

SHIP'S TRACK ⟶

UNCORRECTED FIELD RECORDING ≐ 100 m

A

HORIZONTAL & SLANT RANGE CORRECTED

B

FIGURE 5.7 Sonographs, (A) uncorrected, and (B) corrected for slant range (Klein K-MAPS III), over algal reefs in northern Gulf of Mexico, water depth 70 meters. (Courtesy of Summit Geophysical, Inc., Houston, Tex.)

INTERPRETATION

The side scan sonar technique is analogous to the use of airborne Side Looking Airborne Radar (SLAR) imagery, and permits the detection of morphological and sedimentary variations on the sea floor as a function of acoustic reflectivity. If the sea floor is uniform in both texture and bathymetry, no echoes are recorded on side scan, unless from features in the water column. Disruptions of the sea floor present a target or a

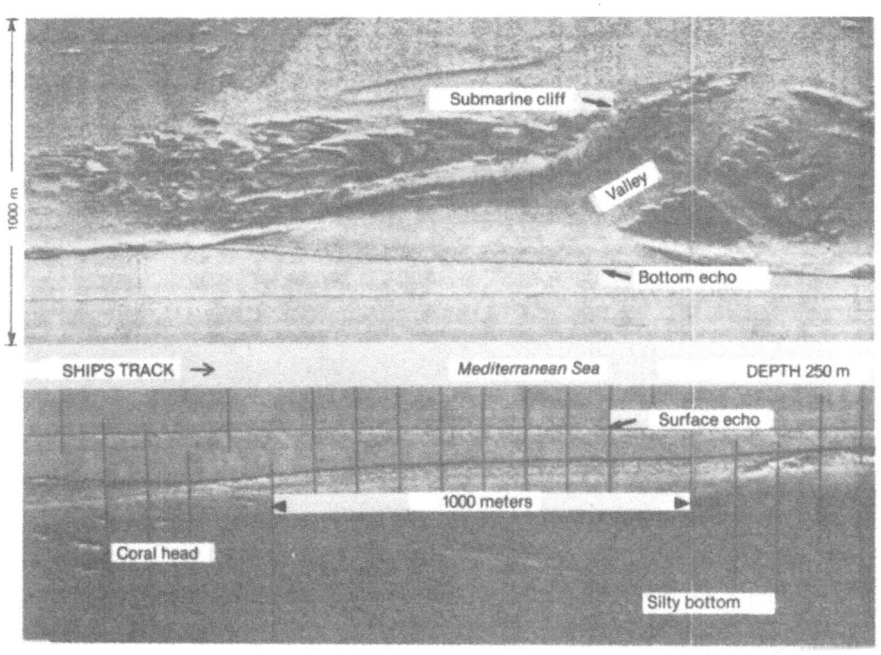

FIGURE 5.8 Sonograph obtained with medium-range side scan sonar system (SOL-120), over submarine valley in Mediterranean. (Courtesy of BEICIP/ RUEIL/MALMAISON.)

reflecting surface displayed as an acoustic reflection. Experience with known objects and features permits accurate interpretations.

On data recorded without slant range correction, ranges marked on the record during the stylus sweep are based upon equal time subdivisions, and thus represent slant ranges. These slant ranges are the direct reflection times between the transducer fish and the sea floor. Ranges must be corrected by use of Pythagorean geometry (assuming a flat bottom) for the slant (fig. 5.7A), in order to provide true distances (fig. 5.7B). Corrective values are available as precomputed nomograms (4) or, in the case of digital recordings, may be done automatically.

The stronger the returned reflection or echo, the darker the recorded mark on the paper recording, depending upon the acoustic reflectivity of the sea floor and topography. For example, larger objects such as boulders, rock pinnacles, ridges, and sand waves, are not only good

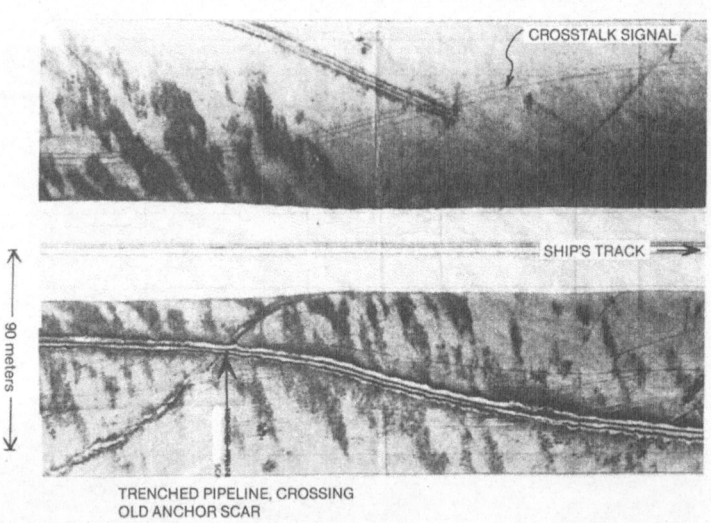

CROSSTALK SIGNAL

SHIP'S TRACK

90 meters

TRENCHED PIPELINE, CROSSING
OLD ANCHOR SCAR

FIGURE 5.9 Side scan sonar depicting pipeline in trench, and crosstalk signal, from Ninian Field, North Sea. (Courtesy of Klein Associates, Inc., Salem, N. H.)

reflectors but also produce large acoustic shadows that leave areas of no return (white shadows) behind the recording away from the transducer. The width of the shadow zone and position of an object relative to the fish may be used to calculate, or at least estimate, the height and size of such objects (4).

The effect of side lobes on sonographs is to produce white gaps between overlaps of the main and side lobes. The resolution within such areas is considerably reduced and occurs close to the inner margins of each channel. Belderson et al. (2) show many excellent examples of sonographs in conjunction with their geological interpretation, providing a most useful reference catalogue.

Common features identified with the use of side scan sonar generated sonographs, which may pose a hazard to the emplacement of structures on the sea floor, include: rock outcrops (linear alignments, fig. 5.8), sand waves and ripples, indicative of bottom currents (linear ridges, fig. 5.1), lateral variations in surface sedimentology (stronger acoustic signals are received from areas of coarser sediments, compared to softer ones, fig. 5.9), zones of sea-floor movement (mud flow gullies and collapse depres-

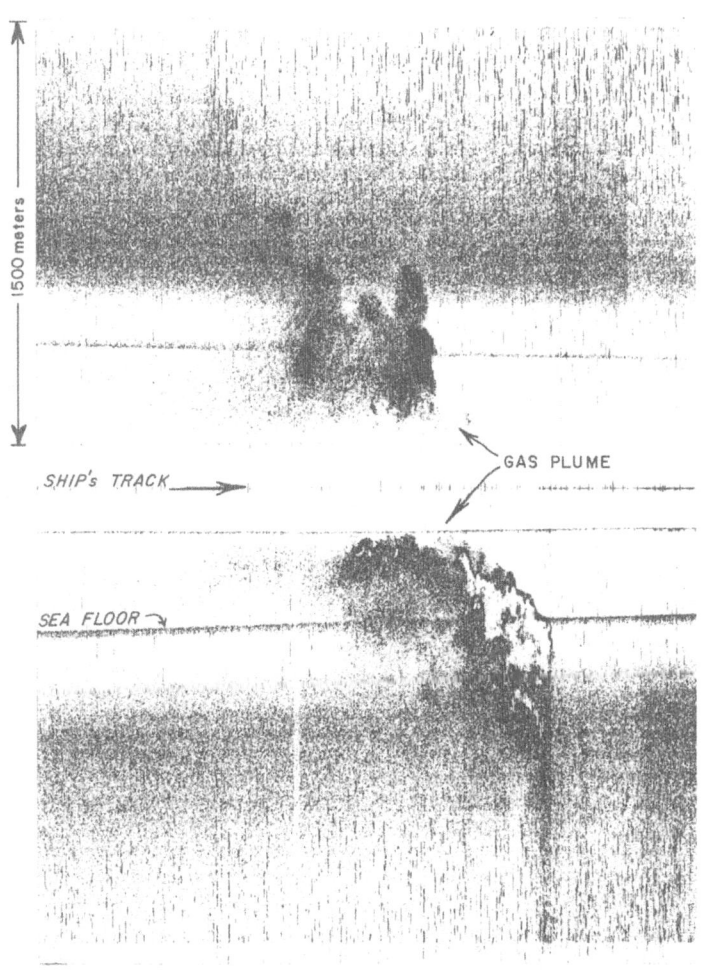

FIGURE 5.10 Sonograph obtained with a medium-range side scan sonar (SOL-120), over a large natural gas seep, northwestern Gulf of Mexico. (Courtesy of Summit Geophysical, Inc., Houston, Tex.)

sions, fig. 5.6), areas of submarine gas seeps (fig. 5.10), areas of sea floor scoured by the passage and grounding of icebergs (fig. 5.11), submarine communication cables, pipelines, submerged wellheads (fig. 5.12), sunken ships (fig. 5.13), and production area debris. Side scan sonar was

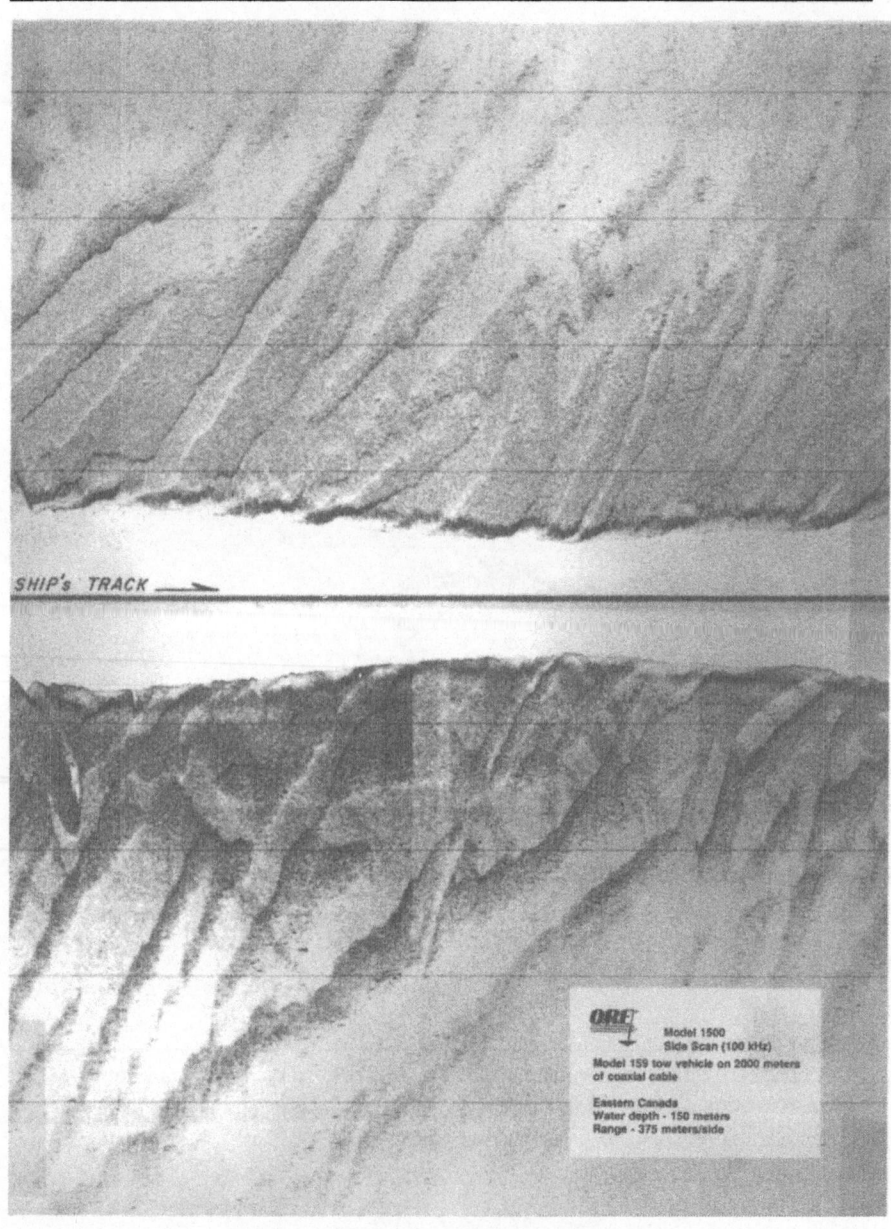

FIGURE 5.11 Sonograph showing iceberg scour marks (incised groves) in sea floor off eastern Canada. (Courtesy of Ferranti–O.R.E., Falmouth, Mass.)

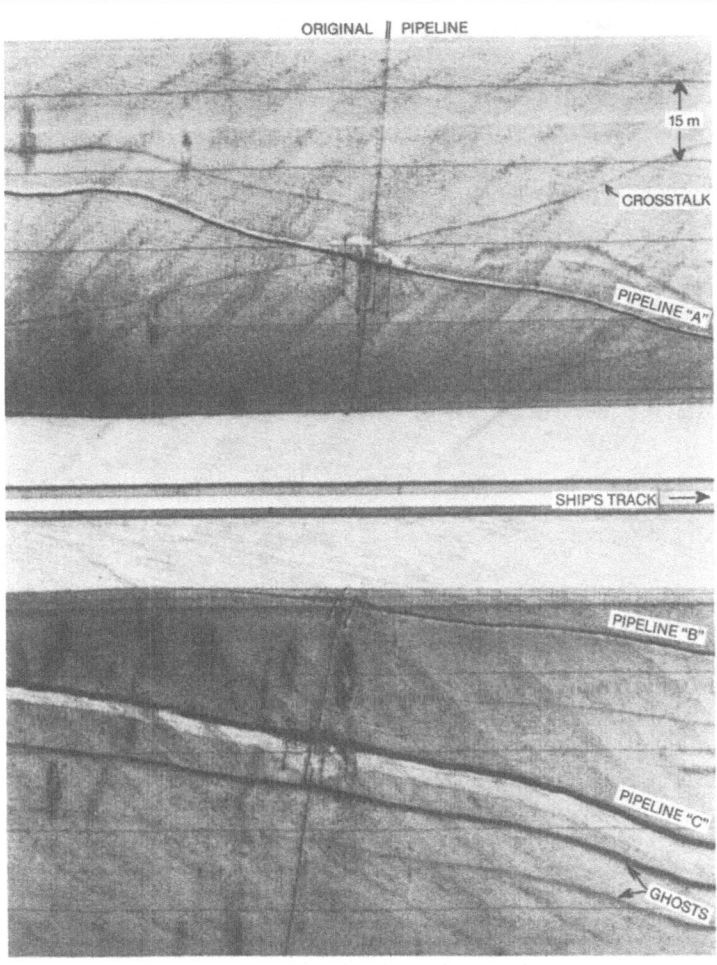

FIGURE 5.12 Sonograph of 3 pipelines (A, B and C), raised on trestles where crossing an originally emplaced pipeline. Arabian Gulf (note ghosts and crosstalk signals). (Courtesy of Klein Associates, Inc., Salem, N. H.)

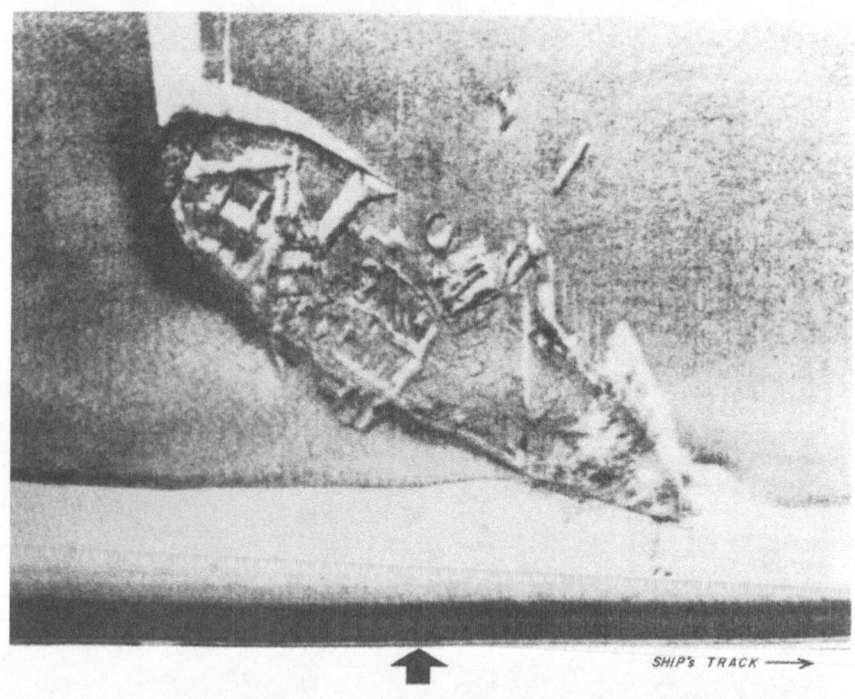

SHIP's TRACK ⟶

FIGURE 5.13 Single-channel sonograph over remains of sunken lightship, Buzzards Bay, Mass. (500 kHz recording, water depth 20 meters). Note scour (erosion) of sea floor about ship's bow. (Courtesy of Klein Associates, Inc., Salem, N. H.)

employed in the search for the *Edingburg*, and the *Crown* (7), and in locating the *Breadalbane* and the *Hamilton*.

State-of-the-art digital side scan sonar systems are apt to be employed more frequently in the future for obtaining detailed bathymetric data, in a mode similar to that of the multibeam echo sounders described in chapter 4. Current developments in the field of side scan sonar data acquisition include synthetic aperture and stereo systems.

Six
SUBBOTTOM PROFILERS

INTRODUCTION

High-resolution seismic reflection systems operating at acoustic frequencies between 1.0 and 14.0 kHz are grouped within the general category of subbottom profilers. These systems allow for continuous high-resolution seismic profile recordings of the uppermost 30 m of strata, below the sea floor.

The subbottom profiler (SBP) has long been a principal tool in the high-resolution investigation of offshore sites for the emplacement of pipelines, bottom-supported structures such as jack-up rigs, and in the exploration for shallow mineral deposits as well as numerous other applications.

HISTORICAL BACKGROUND

The *Marine Sonoprobe*, developed in the early 1950s (1), was the forerunner of present-day subbottom profiling systems. Initial development of the Marine Sonoprobe was aimed at producing a tool for the offshore investigation of recent sediments; however, the system soon found its application in the survey of pipeline routes (1). Numerous systems, which operate at frequencies between 3.5 and 7.0 kHz, are presently available under a variety of trade names.

APPLICATIONS

Subbottom profilers provide continuous seismic reflection profiles in real time at relatively high ship speeds (10–12 kts) and in all water depths, while allowing up to 30 m penetration of the sea floor (2, 3, and 4). The records obtained by these systems reveal in great detail the shallow sedimentary structure of the uppermost strata beneath the sea floor (fig. 6.1). The resolution is typically better than a half meter, depending upon the operating frequency and a variety of other factors.

Subbottom reflection data have been used extensively for pre- and postdredging operations within harbors and channels, at offshore construction sites and for selection of pipeline routes, siting of drilling platforms, in the reconnaissance of sites for the construction of artificial islands for nuclear power plants and petroleum drilling operations, and in the exploration for minerals in shallow waters such as sands and gravels. The continuous seismic subbottom profiles also allow the detection of near-surface hydrocarbon gas seeps within the water column, as shown in the profile recording of figure 6.2. When properly calibrated, SBP data may also be used for obtaining continuous bathymetric profiles.

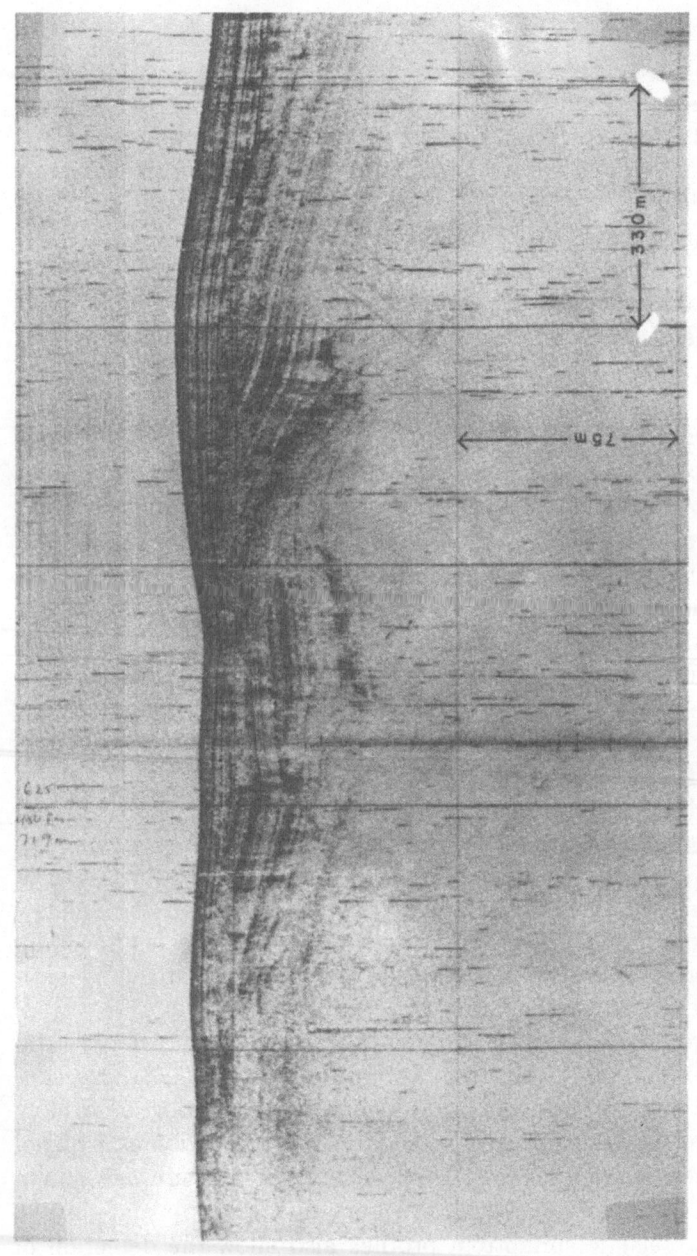

FIGURE 6.1 Typical 3.5-kHz SBP recording from the north Atlantic, water depth 2720 meters. Note unconformity and thrust folds. (Courtesy of Ferranti–O.F.E., Falmouth, Mass.)

83

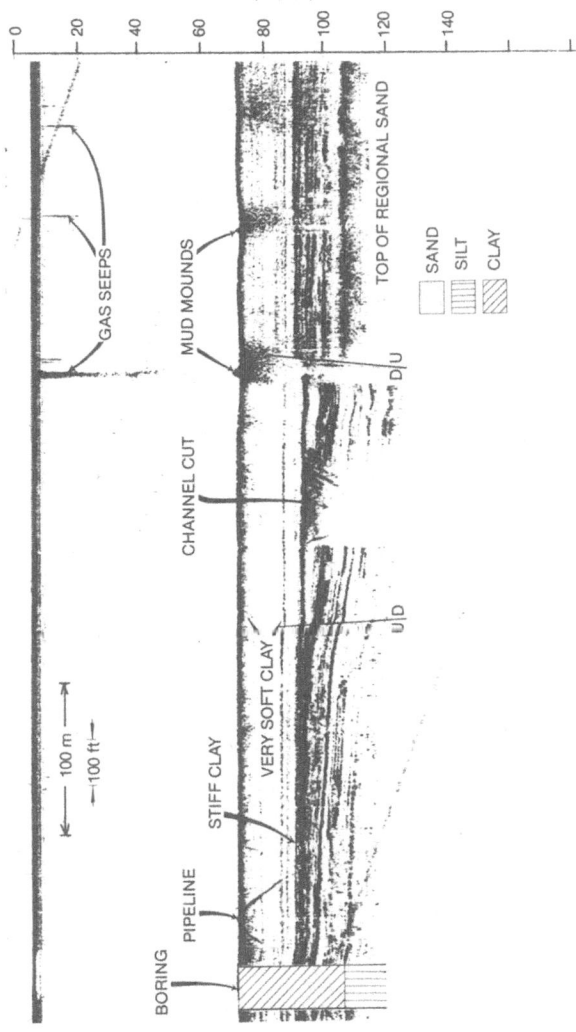

FIGURE 6.2 Subbottom profiler (3.5 kHz) record depicting: geoengineering borehole tie-in, gas seeps, pipeline, mud mounds, faults, and channel cut and fill features, Gulf of Mexico. (From Antoine and Trabant, 1976, by permission of the authors.) (8)

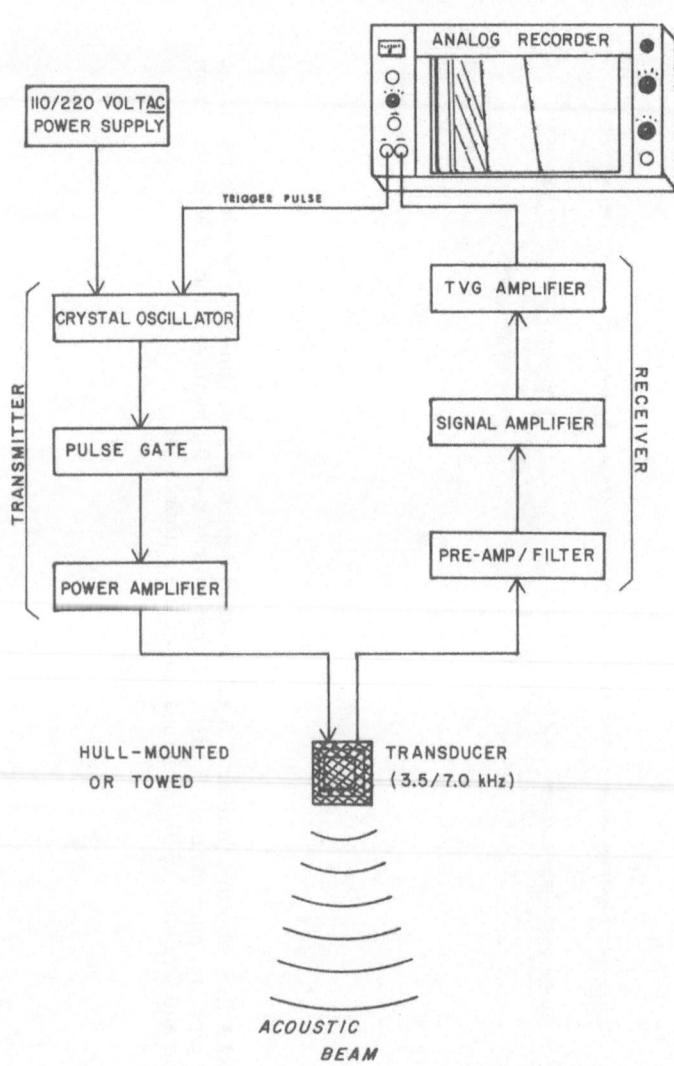

FIGURE 6.3 Electronic block-diagram of subbottom profiler system. Note that all transceiver components are usually housed within a single box. (Diagram by the author.)

PRINCIPLE OF OPERATION

The typical SBP system consists of a transceiver electronics package, a transducer (some units contain a number of transducers for greater power

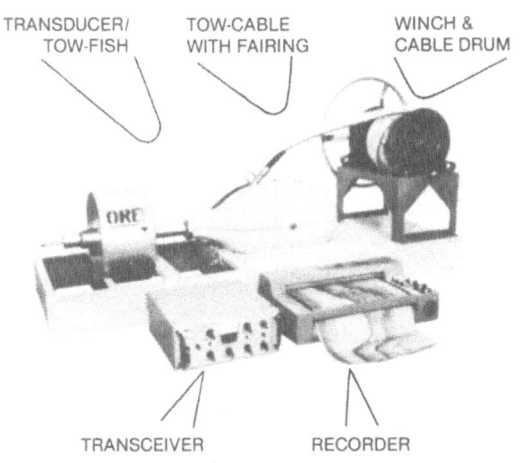

TRANSDUCER/   TOW-CABLE   WINCH &
TOW-FISH   WITH FAIRING   CABLE DRUM

TRANSCEIVER   RECORDER

FIGURE 6.4 Typical subbottom profiler system components: transceiver, winch, transducer fish, and recorder. (Courtesy of Ferranti–O.R.E., Falmouth, Mass.)

output), and a facsimile-type recorder (figs. 6.3, 6.4). The operational circuitry and dimensions are similar to but larger than those of echo sounder systems, in order to operate at lower frequencies (1.0 to 10 kHz), and generate higher power outputs of up to 10 kW. The block diagram of figure 6.3 depicts the basic circuitry common to most systems. Compared to the echo sounder, one finds additional filtering circuitry and time varying gain circuits to boost the received signal amplitude levels with depth of penetration.

Most SBP systems are equipped with electronic circuitry to blank out the water column. While this feature produces a cleaner record, it also masks the presence of features within the water column such as gas seeps, which are of concern to the emplacement and safety of drilling structures.

OPERATION

Continuous SBP records are displayed on facsimile or fiber optic type recorders such as shown in figure 6.4. Such recorders require a variety of operational control features including: trigger signal output for activating the transmitter, signal input from the receiver, sweep and keying time

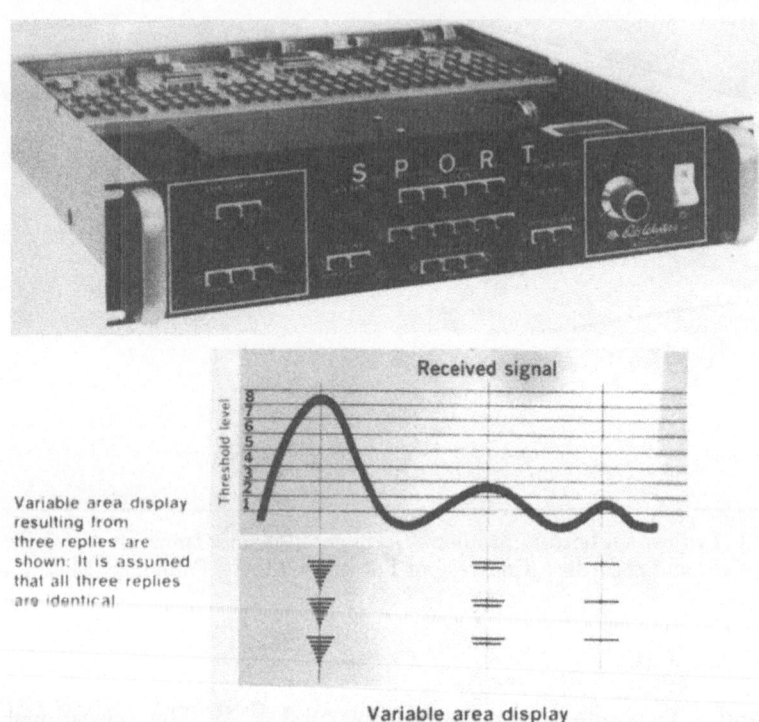

Variable area display
resulting from
three replies are
shown: It is assumed
that all three replies
are identical

Received signal

Variable area display

FIGURE 6.5 Real-time subbottom profiler digital processor system (SPORT). (Courtesy of Edo Western Co., Salt Lake City, Utah.)

controls, scale line settings, and delay circuits. A major requirement for a recorder is that it have a dynamic recording range, gain and sweep speeds to produce a readable record scale, and the varying gray scales associated with the seismic reflectors. The dynamic range for such analog recording systems is generally on the order of 20 to 25 dB. This range can be markedly increased, however, through the use of digital sampling and playback techniques (fig. 6.5).

DIGITAL OPERATION
Digital systems represent the future trend for most electronic instrumentation, and as such, SBP systems have witnessed this development

since the mid-1970s. Conversion of the returned analog reflection data to binary bits allows for temporary storage in memory, and digital microprocessor manipulation prior to graphic display and/or recording on digital tape. Manipulations of the reflection data in this format allow the retention of virtually unlimited dynamic ranges, band pass filtering, time varying gain enhancement, and vertical stacking of adjacent traces for signal-to-noise enhancement (fig. 6.5).

Digital units are available as stand alone transceiver units, or as add-on features for existing analog systems. Digital tape recordings allow for postsurvey processing by seismic array computers for further data enhancement.

OPERATIONAL ADJUSTMENTS

The power output and receiver gain of a subbottom unit in operation needs to be balanced or adjusted to equal levels. This tuning may be performed by listening to the amplified signal with earphones, or observing the two signals on an oscilloscope.

The recordings should provide a clear indication of the first multiple, if present. The purpose of the time varying gain (TVG) circuitry is to boost the gain of the SBP record in order to compensate for signal losses of the acoustic pulse as it propagates through the water column and sea floor. The TVG adjustment should be tuned to bring in an optimum picture of the subbottom layers to a maximum depth. The onset of the TVG may be adjusted manually, to start the gain ramp increase just above the sea floor or, on some systems, placed in an automatic mode whereby the ramp starts with the first return from the sea floor.

If one desires to sample data within the water column, however, the TVG needs to be off or started at time zero. Once these basic settings are made, the acoustic response of the sea floor and underlying sediments becomes the limiting factor toward optimum reception (recording) and penetration.

The transducers required for the operational frequencies of SBP are relatively large. They are commonly mounted within fluid-filled compartments (sea chests) on the hull of the survey vessel, or towed over the side of a survey vessel, within a hydrodynamically streamlined tow body (fish) (fig. 6.4). These practices reduce noise due to water flow, bubble entrapment beneath the transducer, the roll motion of the vessel, and ship-generated noise in general.

Disadvantages of the magnetostrictive and ceramic transducers used for subbottom profilers are their limitations with respect to power output. Overdriving these transducers with excess power results in ringing both at the main operating and related (harmonic) frequencies. This problem is reduced by the use of short transmitted pulse widths and narrow band pass frequency filtering circuitry.

Cavitation is also an occasional problem with subbottom profiling units at higher power outputs (also in heavy seas for hull-mounted units) depending upon the hydrostatic pressure at the transducer. The problem is usually solved by increasing the hydrostatic head by lowering the towed units deeper in the water column (3), or providing pressure to the fluid within the sea chest for hull-mounted units (a water-filled standpipe is usually sufficient).

VARIABLE FREQUENCY UNITS

To date, subbottom profilers have operated in a monofrequency mode, employing a short pulse width at a fixed frequency (i.e., 1 ms pulse at 7 kHz). A number of units are in the process of development, however, which would transmit a longer signal burst containing either several frequencies or consist in a sweep through a portion of the applicable frequency spectrum (5).

Similar developments took place during the 1950s for the development of the Vibroseis (trademark of Continental Oil Company) technique for the exploration of hydrocarbons (6). While the original Vibroseis system used much lower frequencies on the order of 8 to 80 Hz, serious efforts are being made with the higher frequencies employed by the SBP technique. The CHIRP sonar (5) thus employs a pseudorandomly generated signal between 1 and 5 kHz, with individual frequency signal lengths of a few cycles, for a total tone burst length (pulse width) of about 100 ms at 5 kW power output.

The advantages of such multifrequency pulses are that they combine sufficient frequency variability to penetrate most bottoms to optimum depths, without operating separate equipment, while increasing the signal-to-noise ratio. The returned signal must be processed, however, or cross-correlated with the transmitted pulse either in real time, or during postsurvey data processing. This processing must be performed in order to decode the complex returned multiplexed signal, and obtain the optimum resolution or picture of the subsea strata. This latter process

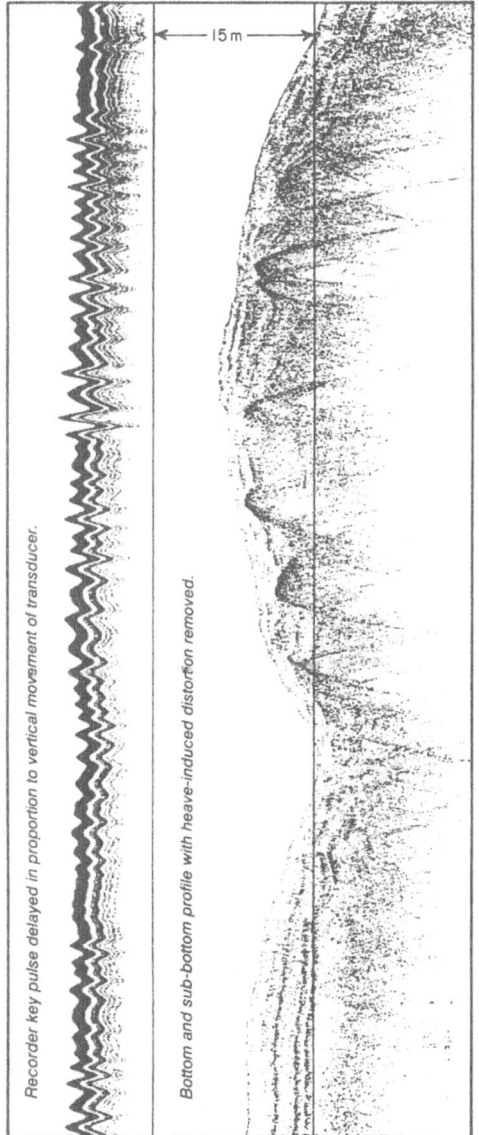

Recorder key pulse delayed in proportion to vertical movement of transducer.

Bottom and sub-bottom profile with heave-induced distortion removed.

15 m

FIGURE 6.6 Subbottom profiler recording (3.5 kHz), illustrating the removal of heave-induced distortion, from Gulf of Mexico. (Courtesy of Ferranti–O.R.E., Falmouth, Mass.)

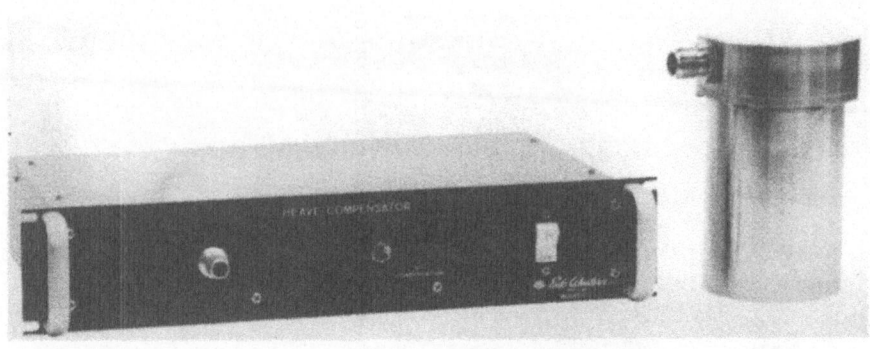

FIGURE 6.7 Heave (ship's motion) compensator processor and accelerometer. (Courtesy of Edo Western Co., Salt Lake City, Utah.)

may be done with ease in real time with the advent of programmable microprocessor systems, and promises great technical advantages within the near future.

HEAVE COMPENSATORS

Several methods have been devised toward removing the effect of survey vessel motion on subbottom profiler recordings. The distortions created by this movement frequently obscure the subbottom reflectors (fig. 6.6), rendering interpretation difficult. The motion of a survey vessel may be cancelled by input from motion sensors (accelerometers, fig. 6.7) or by simply filtering the incoming data with respect to time (swell filtering), and applying a smoothing function relative to the sea floor. This latter technique is relatively simple and produces good results, as shown by the seismic recording of figure 6.6.

DEEP-TOW SYSTEMS

Subbottom profilers have also been successfully operated as part of deep-tow systems. The units are mounted in a tow fish on a cable towed at considerable depth in order to reduce the effects inherent to deep water surveys. These effects include: amplitude loss due to spherical signal spreading, and the effects of side echoes caused by the wide transmission beam (fig. 6.8).

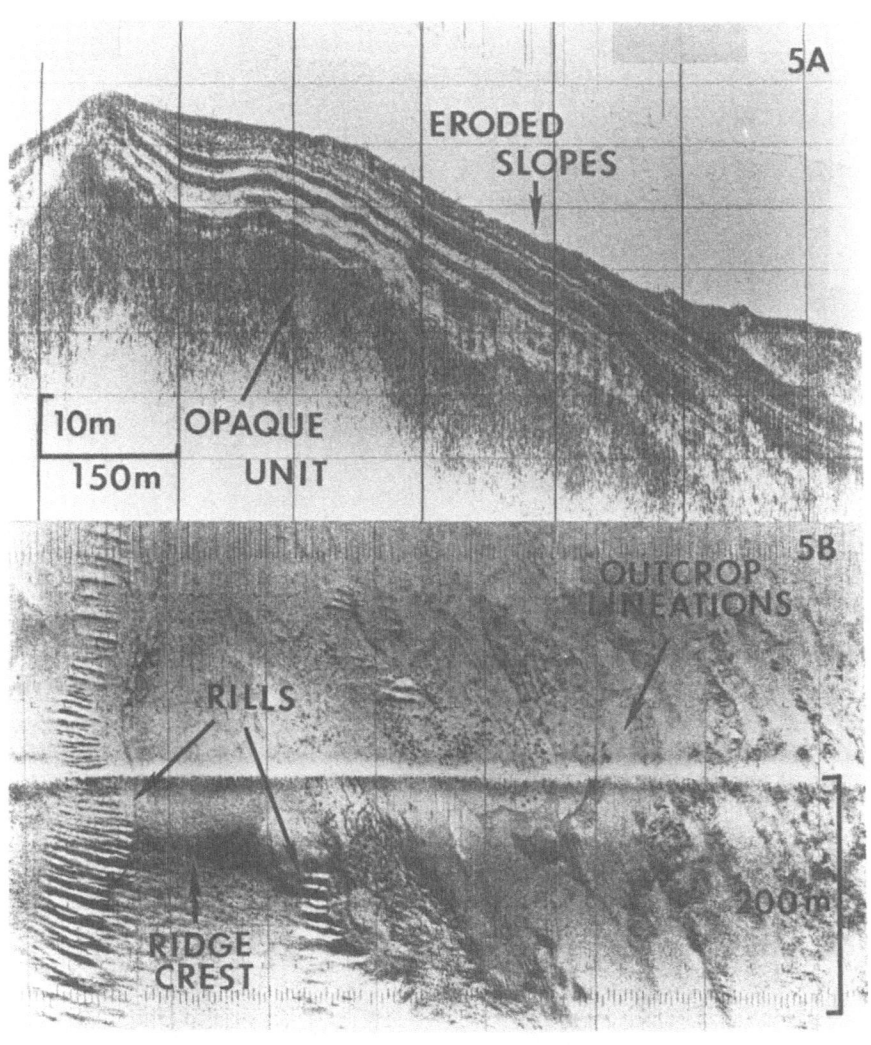

FIGURE 6.8 Matched deep-tow subbottom profile and side scan sonar record (swath width 400 m) of eroded submarine channel ridge flanks, water depth 2650 m, Mississippi Fan, Gulf of Mexico. (From Prior et al. 1983, by permission of the author.) (10)

As deep water operations beyond the shelf break become more common, the use of such deep-tow units is necessary to obtain meaningful high-resolution reflection data.

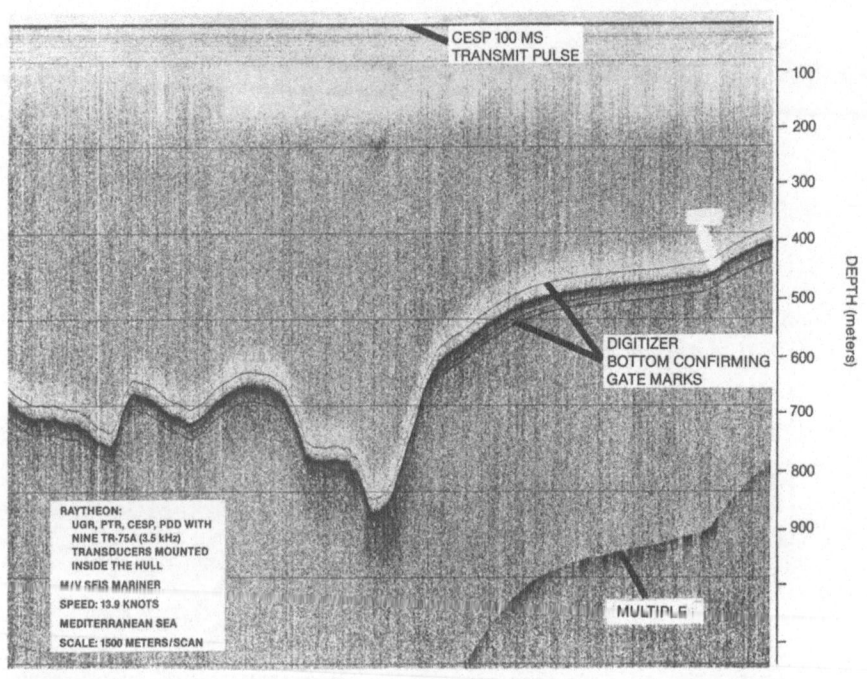

FIGURE 6.9 Subbottom profiler (3.5 kHz) recording, utilizing a signal-correlation processor (CESSP), of acoustically "hard" bottom, Mediterranean Sea. (Courtesy of Raytheon Co., Portsmouth, R. I.)

RESOLUTION

The resolution of recorded subbottom profiles is frequency dependent, as explained in chapter 2. Ideally it should be on the order of 10 cm for 3.5 kHz operation and 5 cm at 7.0 kHz (quarter wavelength at 1500 msec). However, observation has shown the resolving power to be significantly less, perhaps on the order of 1/3 to 1/2 of the wavelength at best (14 to 20 cm for operations at 3.5 kHz). Many variables are involved in the generation of reflectors, particularly the acoustic impedance of the subsea strata.

Correlations of subbottom profiles with shallow borings and core data reveal that very thin layers, such as scattered shell fragments and sand lenses, can produce strong acoustic reflections, while thick layers of

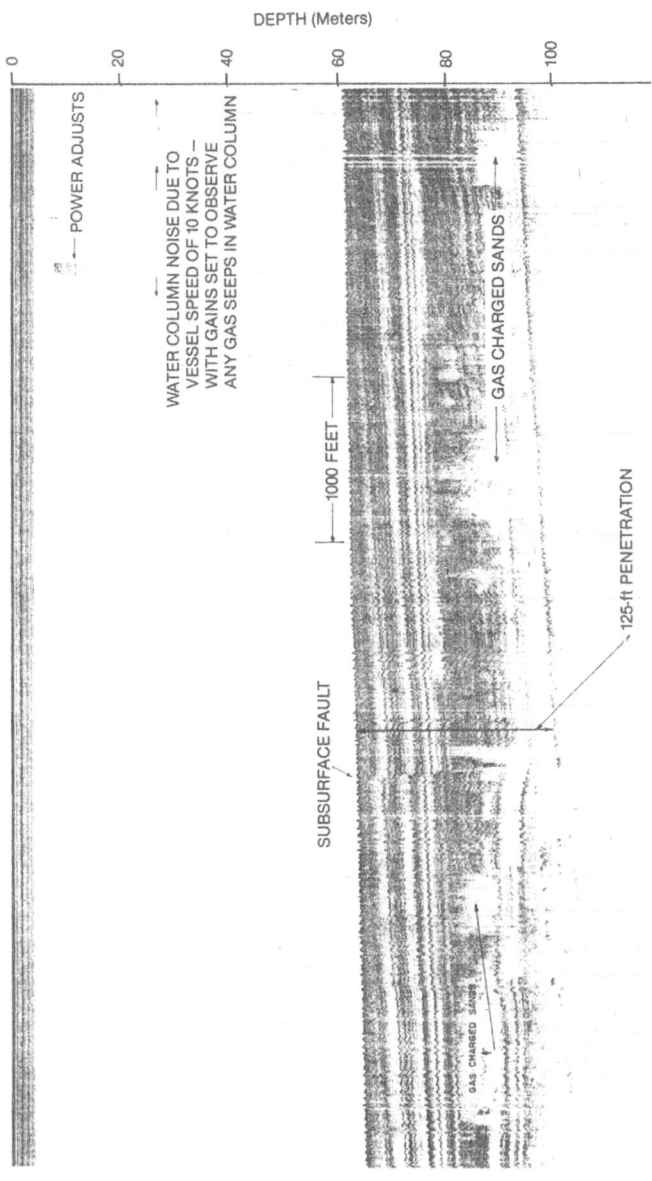

FIGURE 6.10 Subbottom profiler (3.5 kHz) record over acoustically "soft" sediments, displaying over 40 meters' penetration, Gulf of Mexico. (Courtesy of Edo Western Co., Salt Lake City, Utah.)

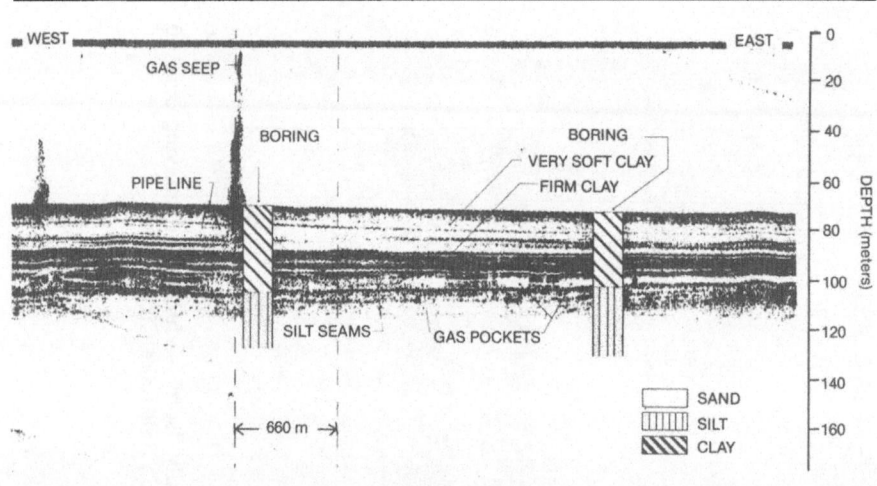

FIGURE 6.11 Example of correlation between subbottom profiler record (3.5 kHz) and geoengineering borehole data, Gulf of Mexico. (From Antoine and Trabant, 1976, by permission of the authors.)

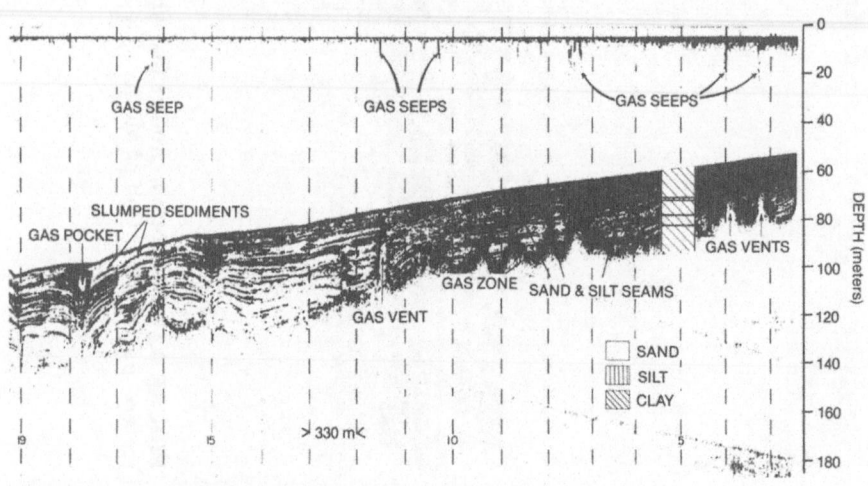

FIGURE 6.12 Subbottom profiler record (3.5 kHz) showing geoengineering borehole data, gas seeps, gas pockets and slumps, Mississippi River Delta, Gulf of Mexico. (From Antoine and Trabant 1976, by permission of the authors.)

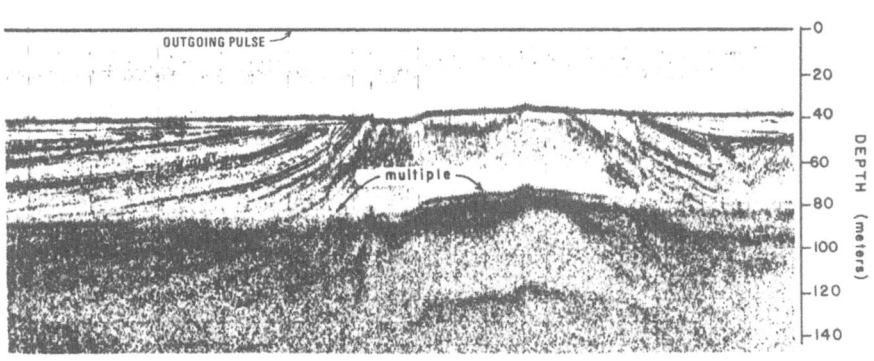

FIGURE 6.13 Example of multiple interference on a subbottom (3.5 kHz) profiler recording, Gulf of Mexico. (Courtesy of Raytheon Company, Portsmouth, R. I.)

relatively different sediment composition may not produce reflections at all. Thus, the composition, physical properties, and microfabric of submarine sediments are important factors, in addition to the recording technique and scale, which will affect the final recorded resolution obtained by a subbottom profiling system.

Penetration of the sea floor, on the other hand, is dependent upon a number of factors, including the operational frequency as well as the sea-floor sediment composition. Examples of these effects are shown on the subbottom profiles of figures 6.9 and 6.10.

INTERPRETATION

Determining the subsea geology from subbottom seismic records, while appearing straightforward at first, is subject to pitfalls that can produce erroneous results. The acoustic reflection characteristics of the sediments, their velocities, and raypaths that result in the seismic profiles, should be well understood before making an interpretation.

Knowledge of the local geology and availability of ground truth in the form of core or boring samples, are invaluable aids in this process. The detailed correlation with engineering borehole data provides an unambiguous interpretation, as illustrated by the profile of figure 6.11.

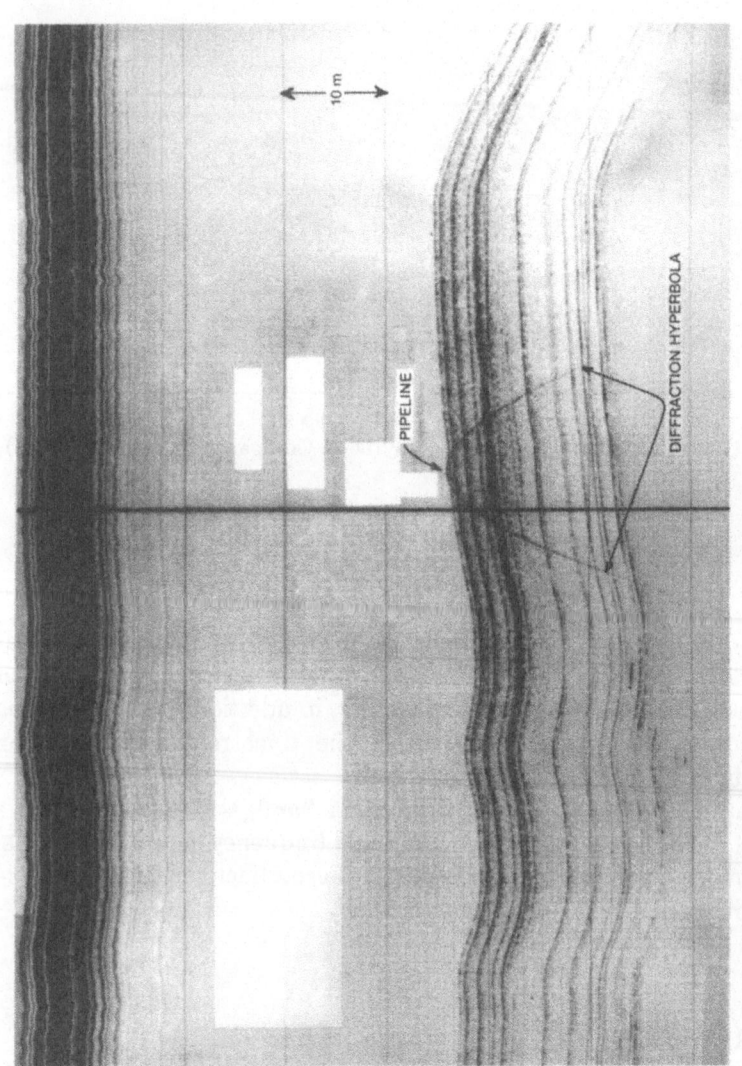

FIGURE 6.14 Subbottom profiler (14 kHz) record, over 30.5 cm (12 in) pipeline in Gulf of Mexico. (Courtesy of Ferranti–O.R.E., Falmouth, Mass.)

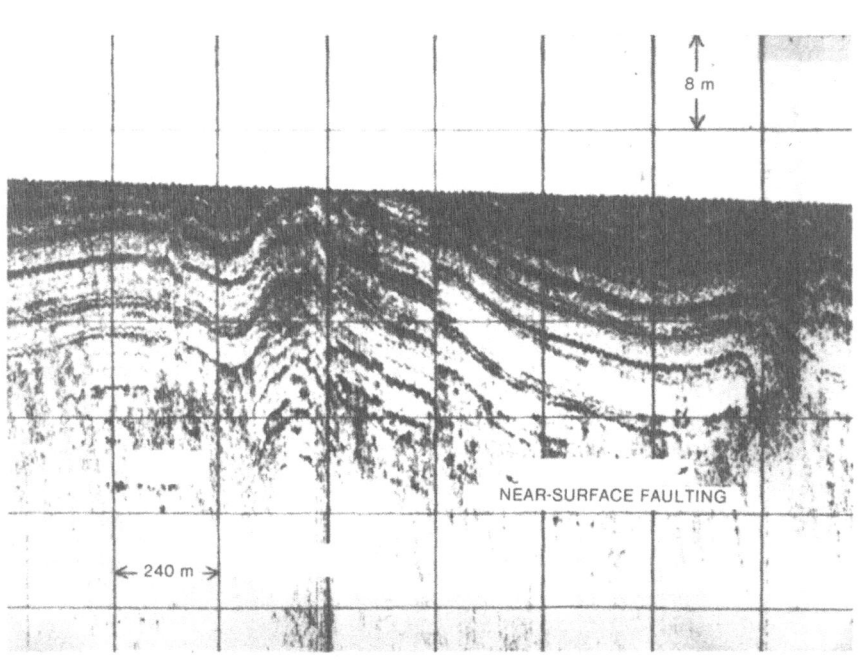

FIGURE 6.15 Subbottom profiler (3.5 kHz) recording showing "active" faulting, and a surface angular-unconformity, Gulf of Mexico. (Courtesy of Fairfield Industries, Houston, Tex.)

Acoustically hard bottoms composed of sand, limestone, or the presence of gas-bearing sediments, may not permit penetration, and hence seismic reflections, as shown in figure 6.9. The use of lower frequencies (chaps. 7–9) may alleviate this problem, while reducing the resolution.

Soft clays, on the other hand, usually allow excellent penetration of the sea floor. Subbottom profilers are commonly referred to as *mud penetrometers* in Europe (7). The seismic profile of figure 6.12 displays that ideal penetration within soft deltaic sediments.

Interference by multiples (chap. 2) within shallow waters frequently limits interpretation to the depth of the first multiple. Strata underlying the first multiple may be masked, as shown by the profile of figure 6.13.

Subbottom seismic profilers are sometimes excellent for locating pipelines and determining their depth of burial (fig. 6.14). The deter-

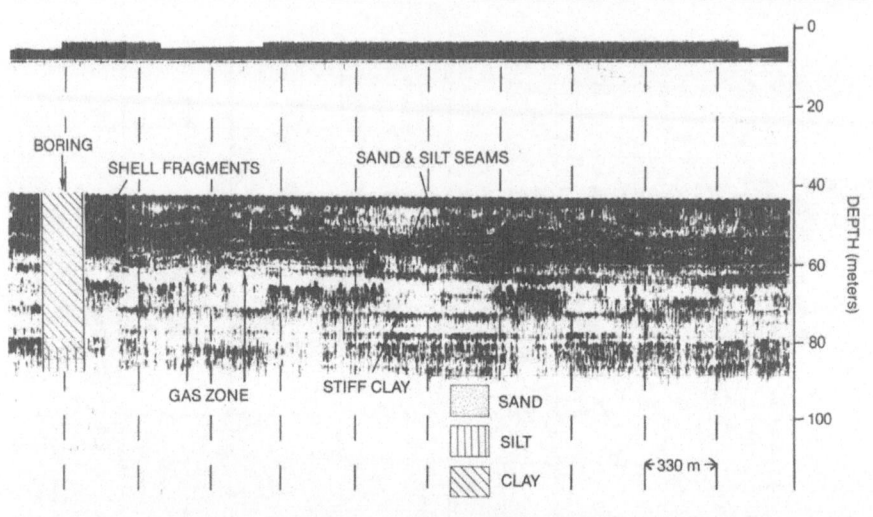

FIGURE 6.16 Subbottom profiler (3.5 kHz) record showing correlation with geoengineering borehole data, and presence of shallow gas-bearing strata. (From Antoine and Trabant 1976, by permission of the authors.)

mination of the orientation and displacement of near-surface faults (fig. 6.15) is relatively simple. The vertical exaggeration of the recordings, however, must be taken into account when establishing dips and slopes.

The degree of activity or movement of faults may also be assessed from SBP profiles according to the amount of near-surface displacement and lack of recent surface sediment cover (fig. 6.15).

The presence of shallow gas-bearing sediments is indicated on SBP records by strong reflections, with underlying wipe-out zones as shown in figure 6.16, and abrupt lateral changes in subbottom reflection characteristics. These lateral changes in subbottom character produce the so-called battleship formations (8), and the presence of shallow gas is occasionally indicated by gas seeps in the water column as well as mud volcanoes (9) or pock marks (mounds and depressions) on the sea floor.

Relict features such as infilled river beds, stream channels, and canyons are easily identified on subbottom profile records, as revealed by the cut and fill strata of figure 6.17. Such features are usually the result of glacio-eustatic (worldwide) fluctuations in sea level, associated with the waxing and waning of continental ice sheets during the Pleistocene Epoch. The effect of these transgressive/regressive cycles was to alternately expose

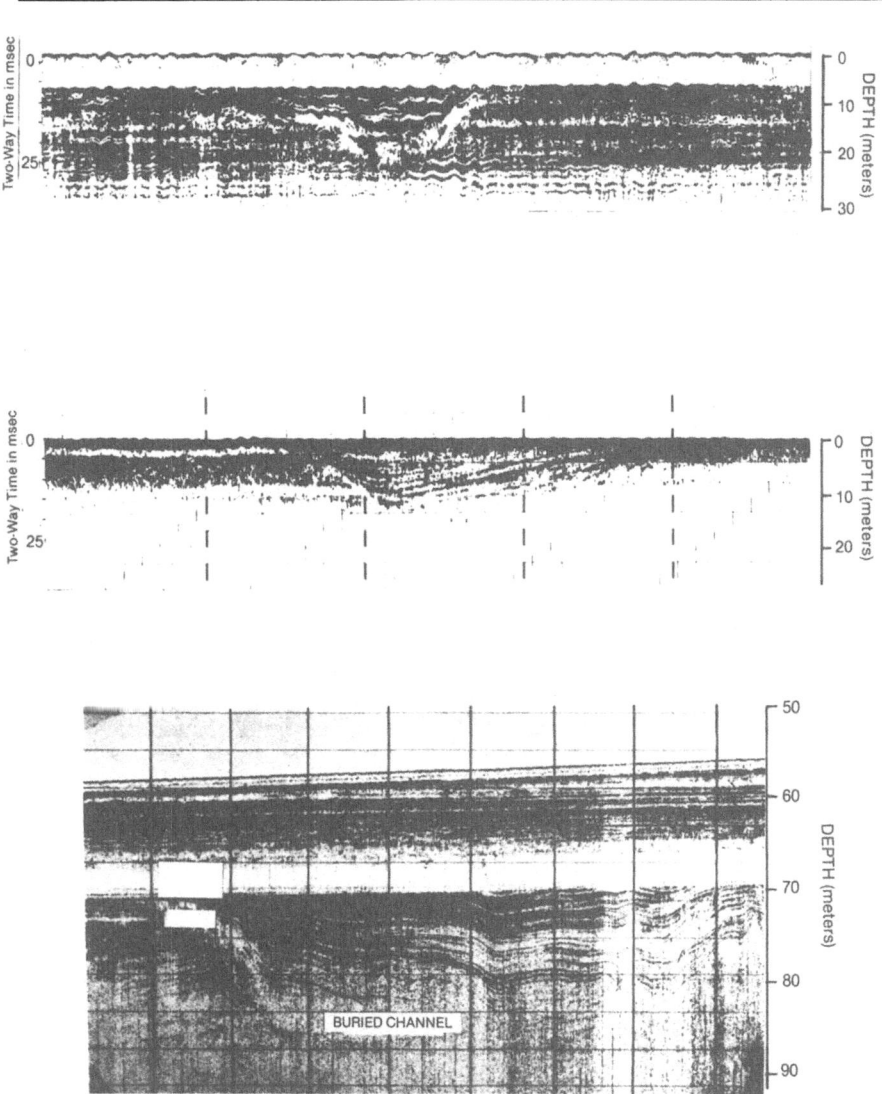

FIGURE 6.17 Subbottom profiles (3.5 kHz) depicting relict infilled river chan-nels, or "cut and fill" features. (Courtesy of Fairfield Industries, Houston, Tex.)

and inundate areas of the present continental shelf to depths on the order of 150 m below present sea level. The curve in figure 6.18 depicts the relative changes in sea level since the last major glaciation.

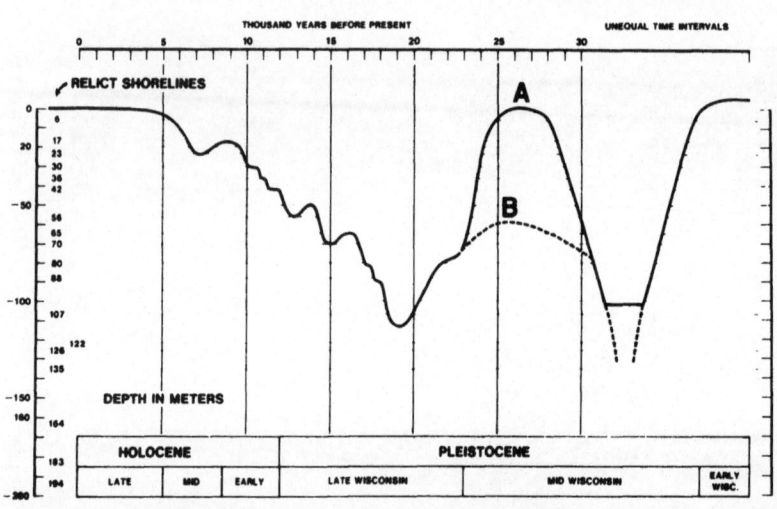

FIGURE 6.18 Curve of relative change in sea level for the northern Gulf of Mexico. Curve derived from various sources. (Courtesy of Coastal Environments, Inc., Baton Rouge, La.)

In summary, the subbottom profiler provides a most important tool in the conduct of HRG surveys, as it establishes the geology and shallow seismic stratigraphy immediately beneath the sea floor.

Seven
SEISMIC SOURCES

INTRODUCTION

It is necessary to generate a fairly strong seismic pulse, in order to obtain subsurface penetration in excess of 1000 m, within the high-resolution frequencies (band pass centered about 100 Hz). There are currently a number of seismic sources that meet these requirements.

Explosive charges had served as the deep penetrating seismic source until the late 1950s, when they were replaced for environmental and safety reasons. The conduct of offshore geophysical surveys then witnessed the development of alternate energy sources which allowed continuous detonations or bangs, with fewer problems than those associated with the explosive charges.

The term continuous seismic profiles (CSP) became prevalent during the 1960s for the acquisition of single channel reflection data. Sparkers were a principal sound source in wide usage by both oceanographic institutions and the petroleum exploration industry during this period.

A number of new and innovative seismic sources for the acquisition of high-resolution seismic reflection data have been developed since the early 1960s. A listing of the most commonly employed sources, and a comparison of individual characteristics, is presented in table 7.1. A graphic presentation of the frequency, power output, and typical subsurface penetration for these sources is illustrated in figure 7.1.

An ideal seismic source should produce an instantaneous positive shock wave, characterized by a very quick rise in pressure, followed by a rapid decay. This requirement for a clean seismic pulse has received a good deal of attention in the design and development of modern seismic sources.

The following sources are the most prevalent in use today, and the discussion shall be limited to them. The first two sources (sparker and air gun), are widely used for the acquisition of CSP data, while the last three systems discussed (water gun, miniflexichoc and minisleeve exploder) represent a relatively new generation of seismic sources employed toward the field acquisition of HRG seismic reflection data. These systems represent the present state of the art for the recording of multifold seismic data, while providing the necessary acoustic power to produce penetrations of the sea floor on the order of 1.0 second (1,000 m) and more.

SPARKERS

For the past several decades sparker sources, which operate by discharging a high-voltage electrical current between electrodes towed at shallow depths, have seen widespread usage in the acquisition of high-

	BOOMER™	1KJ SPARKER	4.6KJ SPARKER	8KJ SPARKER	15.4KJ SPARKER	FLEXICHOC™	MINISLEEVE EXPLODER™	15in³ WATERGUN™	40in³ AIR GUN / 60in³ WATER GUN
TECHNICAL SPECIFICATIONS									
MANUFACTURER	EG&G	EG&G	EG&G	FAIRFIELD	I.F.P	VARIOUS	S.S.I.	BOLT	E.E.S.I.
MODEL & N°	230 & 234	267 A B / 231A, 232A	402/7, 231A / 2x 232A, 2x233A	SS 75	FHC 50	ENGINEERING PROFILER	MICRO-MICA "T"	PAR 800	HYDROSHOCK
SOURCE DIM.	158x84x159 cm	25x25x270cm	33x33x275cm	100x60x60 cm	65x30x30cm	30x5x5 cm each	56x15x15 cm	45x10x10 cm	90x10x10 cm
SOURCE WT.	90 kg	8 kg	10 kg	25 kg	10 kg	30 kg	15 kg	19 kg	20 kg
ENERGY SOURCE DIM	51x48x56 cm	110x60x78 cm	161x113x143 cm	212x146x208 cm	280x110x185 cm	170x105x180cm	200x60x60 cm	200x60x60 cm	200x200x100 cm
ENERGY SOURCE WT.	73 kg	187 kg	263 kg	369 kg	600 kg	620 kg	200 l / 200 kg	200 l / 200 kg	450 kg
TOW SYSTEM	CATAMARAN DEEP TOW HULL MOUNT	3 ELECTRODE TOW SLED	9 ELECTRODE TOW SLED	10 ELECTRODE TOW CABLE	CATAMARAN	12 GUN 800 kg CATAMARAN	CATAMARAN TOW CABLE	TOW CABLE	TOW CABLE
NOTES	5kW, 240V supply	15kW, 240V supply	15kW, 240V supply	30kW, 240V supply		gas cylinders		WAVE SHAPING KIT AVAILABLE	
ENERGY	MAGNETO-STRICTIVE PLATE	ELECTRICAL CAPACITIVE DISCHARGE →	→	→	PLATE EXPANDED HYDRAULIC RE-TRACTED IN VOID	EXPLOSION within ELASTIC SLEEVE	IMPLOSION of WATER SLUG	IMPLOSION of AIR BUBBLE	IMPLOSION of WATER SLUG
SOURCE PERFORMANCE									
FIRING RATE	0.17 sec	0.75 sec	2 sec	2 sec	2 sec	300 msec sequentially, 2 sec otherwise	0.5 sec	0.25 sec	1 sec
PULSE	400-14,000 Hz / 107 m bar m	200-5,000 Hz	70-1,500 Hz	40-900 Hz / 1.0 bar m	30-1,200 Hz / 3.0 bar m	50-1,000 Hz / 5.0 bar m	50-1,000 Hz / 3.8 bar m	40-450 Hz / 1.25 bar m	40-450 Hz / 1.3 bar m
PENETRATION	40 m	150 m	450 m	1,200 m	3,500 m	1,500 m	1,500 m	1,500 m	1,500 m
M.T.B.F	≈ 100 h	≈ 100 h	≈ 100 h	≈ 100 h		6 h (maint)	6 h (maint)	6 h (maint)	6 h (maint)
NOTES	ELECTRODES REQUIRE MAINTENANCE EVERY FEW HOURS →		→	→		REPLACE RUBBER MEMBRANE EVERY 30 HRS	"0" RING REPLACEMENT →	→	

TABLE 7.1 Approximate specifications for a number of HRG seismic sources. Source: manufacturers' brochures.

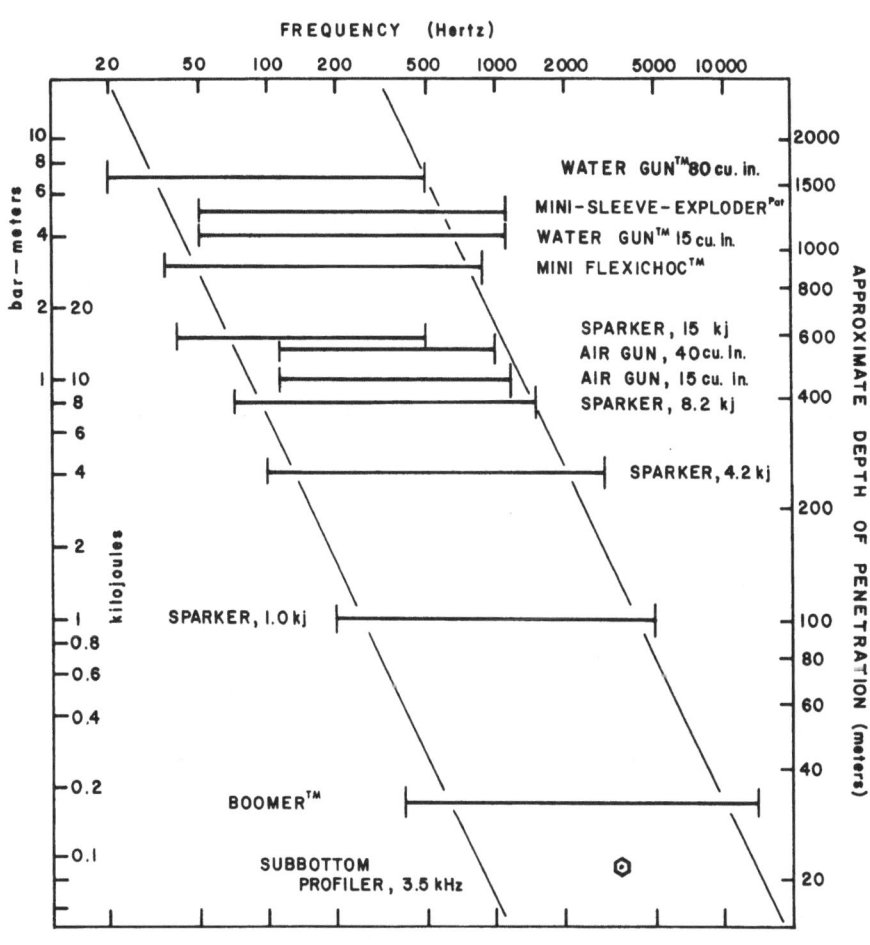

FIGURE 7.1 Graphical comparison of a number of HRG seismic source parameters, frequency, power, and penetration. (Diagram by the author.)

resolution seismic reflection profiles. Figure 7.2 depicts the principal components of a typical sparker system. Sparkers operating in the 1,000 to 20,000 joule power range provide a relatively simple and cost effective seismic source.

Seismic signatures generated by sparkers, however, leave much to be desired. The combination of variable discharge paths and the generation of bubble pulses (chap. 2) tends to produce a relatively dirty or complex

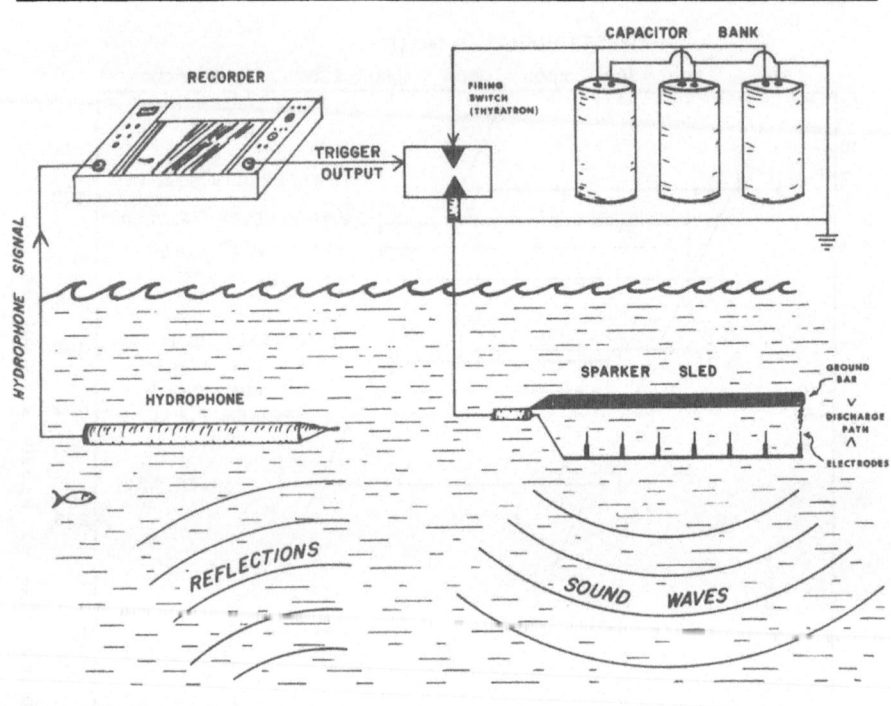

FIGURE 7.2 Schematic diagram of typical sparker system. (Diagram by the author.)

signature (fig. 7.3). This is particularly the case when employed for single channel analog acquisition, and when used in shallow (less than 50 m) water depths (1). It is not uncommon to generate signatures, including the bubble pulse train, of 30 to 50 msec in length, depending upon the source depth. This effect results in the masking of seismic reflectors immediately below the sea floor and beneath subsequent reflecting horizons (figs. 7.4, 7.5). Thus, both resolution and identification of reflectors within 30 to 50 msec below prominent reflectors are masked on single channel reflection data.

On the other hand, the complex sparker source signal becomes an asset when recorded digitally in conjunction with wide range dynamic gain (e.g., floating point). Such near-source recording allows the signature to be correlated or deconvolved with the recorded seismic reflections during processing, and produces a narrow signature or spike through a process of inverse filtering. Such processing techniques, when employed on either

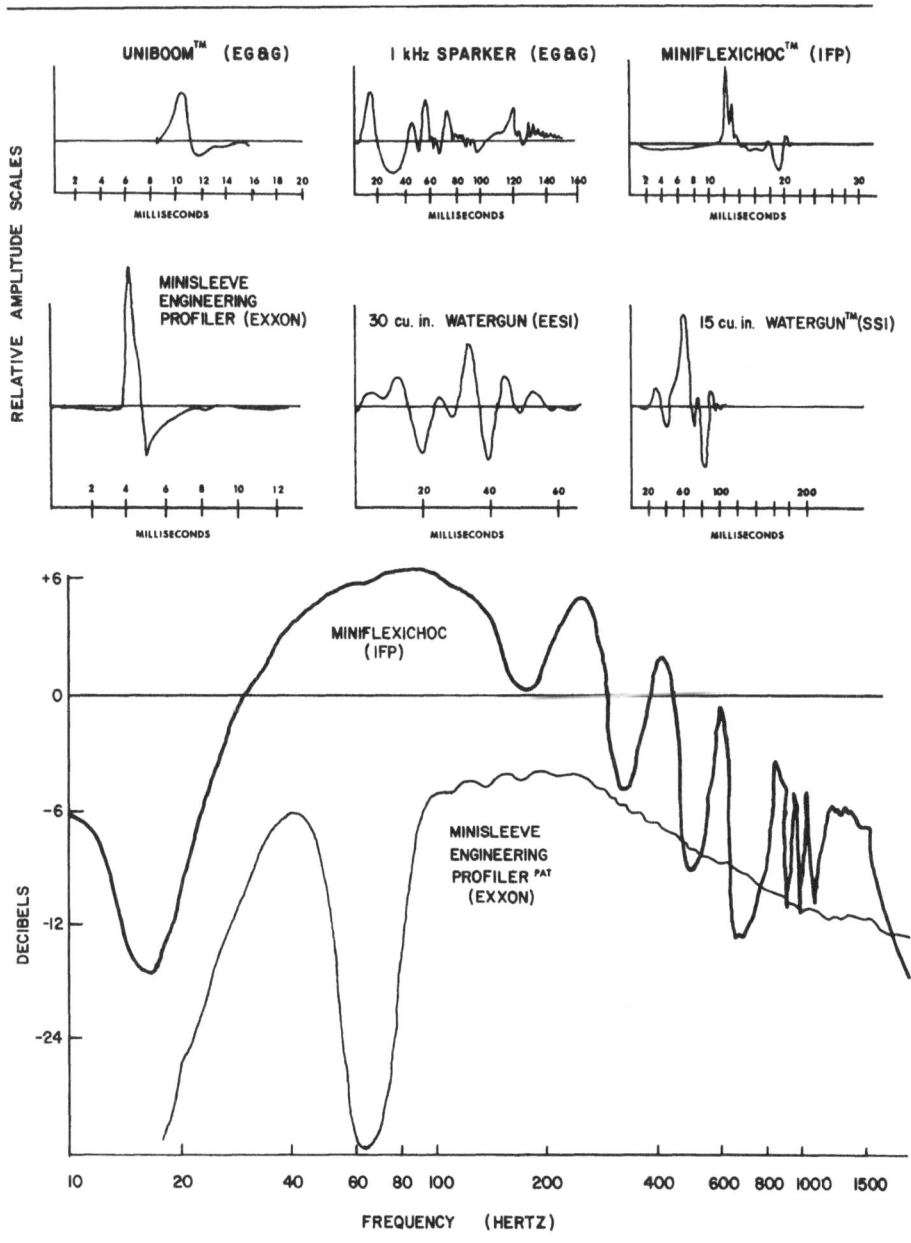

FIGURE 7.3 Signatures and frequency response curves for a number of HRG seismic sources. (Compiled by the author from manufacturers' brochures.)

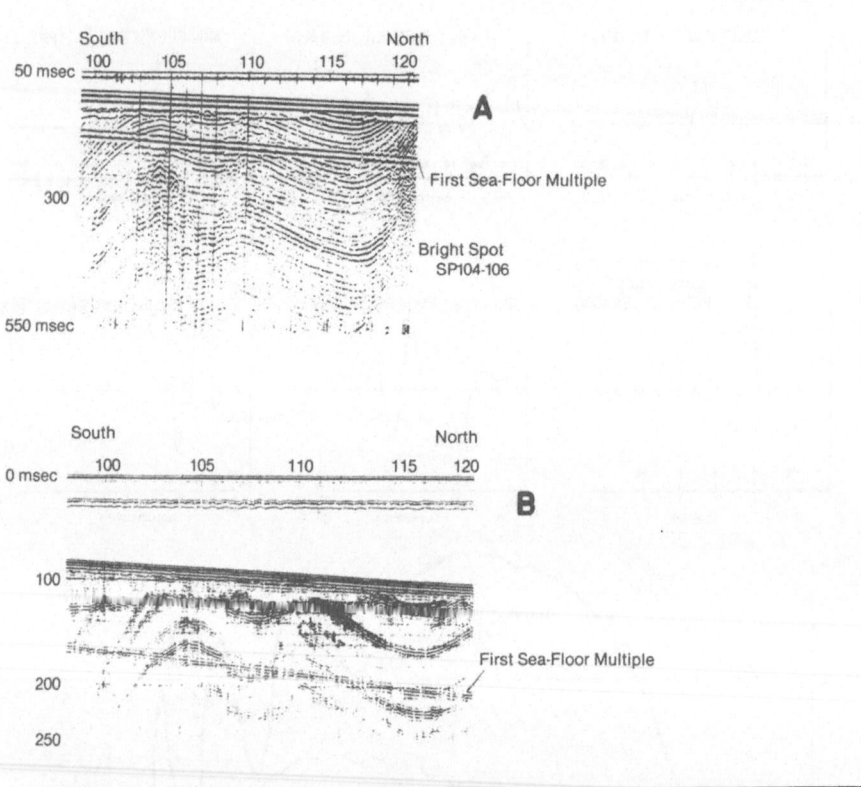

FIGURE 7.4 Comparison of two HRG sparker profiles—(A) 9.0 and (B) 4.5 kilojoules—from Santa Maria Basin, California. Note multiples. (Courtesy of Nekton, Inc., San Diego, Calif.)

single channel or multifold data, permit considerable enhancement by reducing the effects of reverberation, ghosting, and diffraction signals. A reduction in bubble size as well as the resultant ghosting effect may also be obtained by increasing the number and array length of the discharging sparker electrodes. This technique has led to the development of multielectrode sparkers (MES) for HRG profiling, where up to several hundred electrodes may be used to generate the seismic signal pulse.

General specifications for several sizes of sparker sources used for high-resolution seismic surveys are listed in table 7.1.

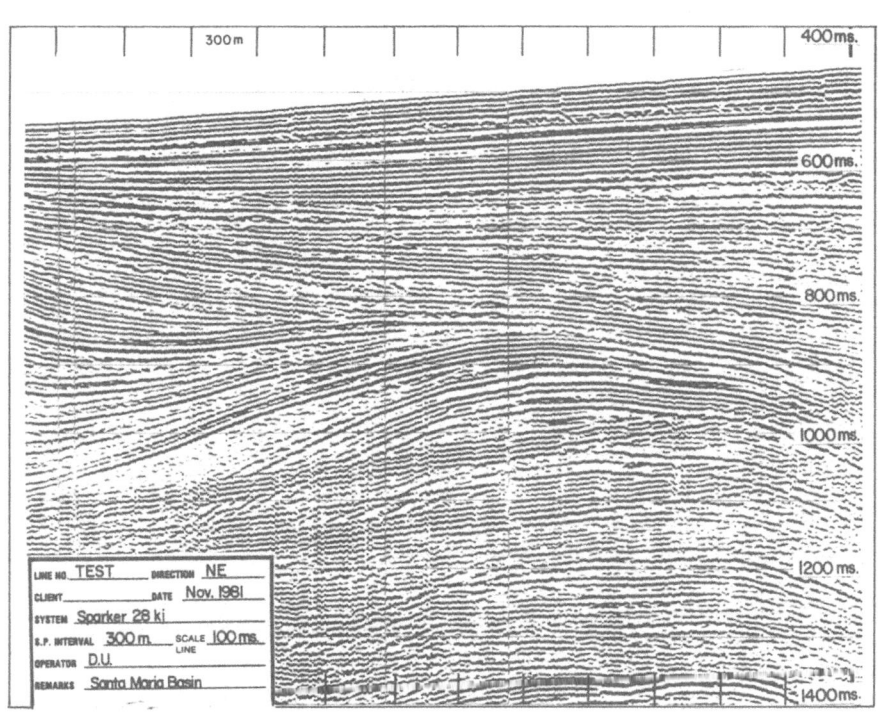

FIGURE 7.5 Single channel HRG analog profile, obtained with 28-kilojoule sparker, Santa Maria Basin, California. Note multiples at 950 msec. (Courtesy of Nekton, Inc., San Diego, Calif.)

AIR GUNS

The use of air gun seismic sources for HRG and CSP surveys became widespread during the 1960s, and is still the major source in use for the acquisition of petroleum exploration reflection data.

While improving the source signature toward a spike, compared to that produced by a sparker, small-chambered (less than 200 cu. in.) air guns tend to generate a strong bubble pulse train, with the deleterious effects of both ghosting and reverberation. The problem is somewhat alleviated by the use of wave shaping devices that break up the secondary bubble oscillations. Also the towing of air guns at very shallow depths of a meter or less has been helpful toward producing a sharper acoustic pulse.

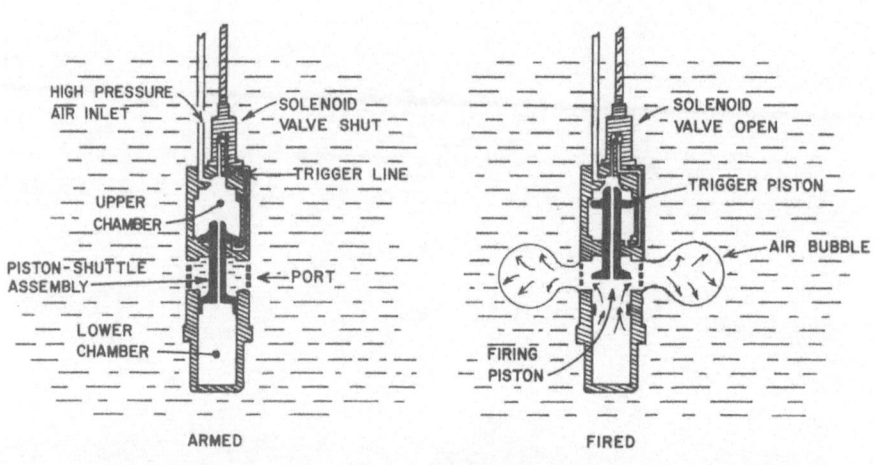

FIGURE 7.6 Diagram of PAR air gun construction and operation. (Courtesy of Bolt Technology Corporation, Norwalk, Conn.)

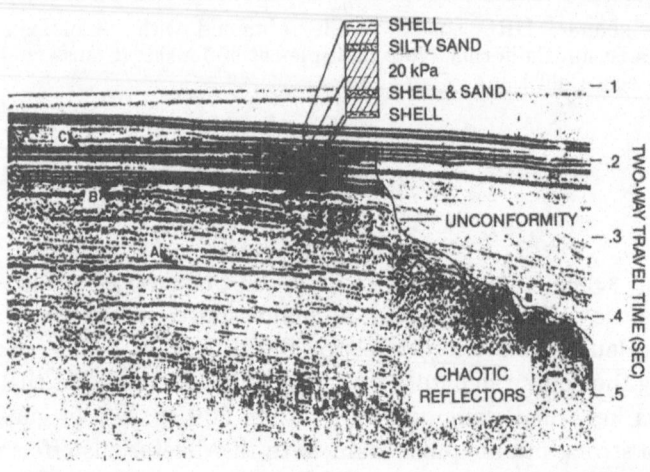

FIGURE 7.7 Single channel HRG profile, obtained with 10 cubic inch air gun, Mississippi Delta, Gulf of Mexico. (From Trabant and Bryant 1978, by permission of the authors.)

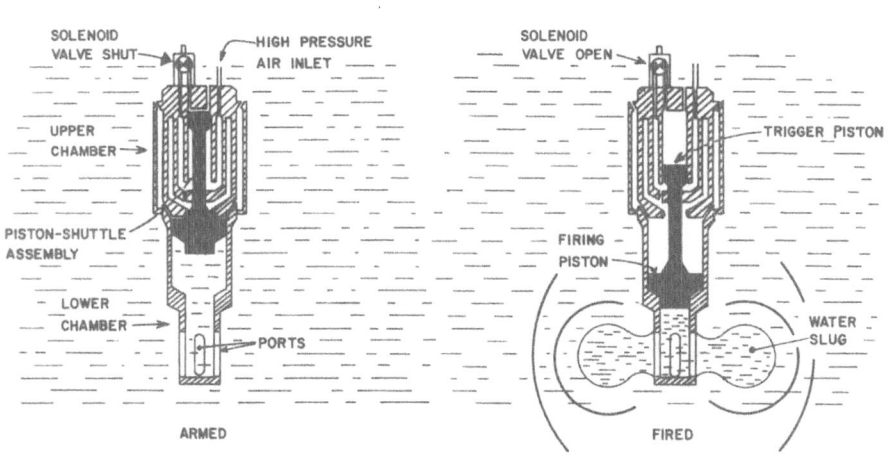

FIGURE 7.8 Diagram of water gun construction and operation. (Courtesy of Seismic Systems, Inc., Houston, Tex.)

The principle of operation, depicted in figure 7.6, consists in the filling of a fixed volume cylinder (e.g., 20 cu. in., or 328 cc) with compressed air (pressure usually between 1000 to 5000 *psi*, or 70 to 352 kg/cm^2). Upon firing, the gun submerged in water, the compressed air is released into the water column to form a bubble proportional to the volume and pressure of the air, and the water depth of the gun.

Large air gun arrays, totaling up to several dozen units, with total chamber volumes on the order of 2000 cu in. have served as a major seismic source in exploration geophysics for more than a decade. Smaller units (less than 200 cu in.) and arrays have been used extensively for high-resolution seismic surveys (fig. 7.7).

Although receiving considerable use in petroleum exploration surveys, small-chambered high pressure (5000 psi) air guns have rarely been employed for high-resolution geoengineering surveys.

WATER GUNS

In an attempt to reduce some of the drawbacks associated with the sources described above, and in order to obtain a cleaner seismic pulse, the water gun was developed in France by SODERA (Societe pour le Developpement de la Recherche Appliqué).

Water guns employ some of the basic mechanical principles of the air gun, except that instead of evacuating a volume of high-pressured air they rapidly expel a fixed volume of water. The outward moving slug of water forms a cavitation pocket that implodes, forming a sharp seismic pulse (fig. 7.8).

One drawback of this system is the generation of a low-frequency pulse prior to the initial cavitation collapse, which is caused by the initial evacuation of the gun chamber. This portion of the signature pulse (fig. 7.3) makes it difficult to determine the initial time break, a very critical element in the processing of multifold HRG seismic data.

Water guns are operated in arrays of 16 to 20 units of 80 cu in. each in petroleum exploration surveys, whereas smaller 15 cu in. units (most commonly two or four guns) are deployed at shallow depth for HRG seismic surveys, as shown by the recording of figure 7.9. Additionally, the use of single 80 and 160 cu in. units has also proven effective as a high-frequency seismic source.

MINIFLEXICHOC

Another development for obtaining an ideal signature for HRG sources, by I.F.P. (Institut Français des Petroles), is the miniflexichoc. This source generates a sharp seismic pulse through the rapid retraction of a flexible metal plate. The unit is subsequently rearmed by a hydraulic jacking system that pushes out the plate. The mechanics of this operation are illustrated in figure 7.10.

The ease of operation of hydraulics, the ability to operate at repetition rates of up to .25 sec, and the sharp signature (fig. 7.3) make the system very effective as a source for HRG seismic surveys. Figure 7.11 depicts a twelvefold miniflexichoc reflection profile off the coast of Venezuela.

MINISLEEVE EXPLODER

The sleeve exploder was developed by Exxon Production Research Co. Scaled down for use as a HRG seismic source, the system consists of one to two dozen rubber membrane sleeves, each of which is fitted around a mixing/firing chamber mounted under a tow sled (fig. 7.12). A mixture of oxygen and propane is fed into the firing chambers and ignited by a spark to produce an explosion contained within each of the rubber sleeve membranes. Upon collapse of the initial explosion, the elastic membranes retract, allowing the exploded gasses to be evacuated through a tube to the surface, thus eliminating the generation of a bubble pulse.

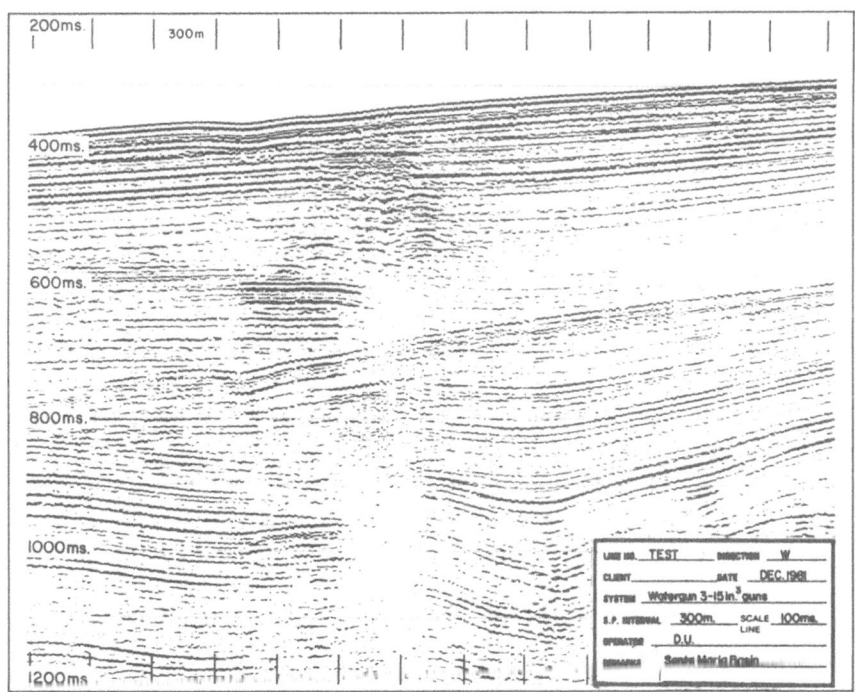

FIGURE 7.9 Single channel HRG profile, obtained with array of 3 water guns (15 cubic inch chambers), Santa Maria Basin, California. (Note prominent bright spot at 600 msec.) (Courtesy of Nekton, Inc., San Diego, Calif.)

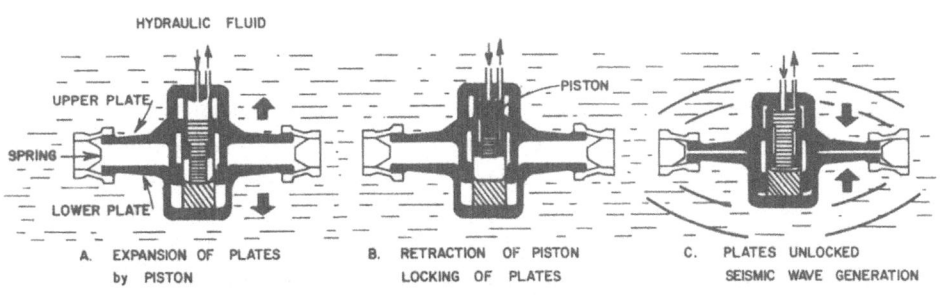

FIGURE 7.10 Diagram (simplified) of Mini Flexichoc construction and operation. (Courtesy of BEICIP/RUEIL/MALMAISON.)

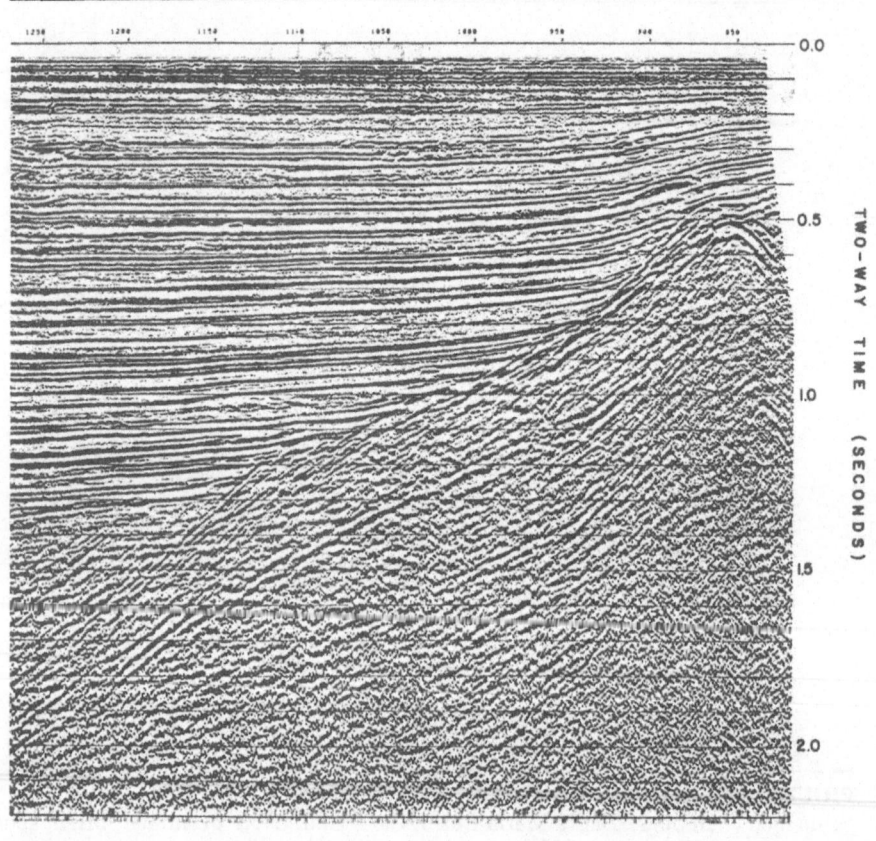

FIGURE 7.11 Processed multifold HRG profile, obtained with Mini Flexichoc source, Venezuela. (Courtesy of BEICIP/CGG/INTEVEP.)

The system is usually operated at shallow depths, which virtually eliminates ghost reflections. The major drawbacks are the size, weight, and use of explosive gasses. In spite of these drawbacks, the minisleeve exploder has gained widespread use for HRG geoengineering surveys designed for the detection of potential hazards to offshore drilling operations, such as faulting and high-pressure gas zones. The seismic profile of figure 7.13 shows comparison with a sparker for a twenty-four-fold seismic reflection profile.

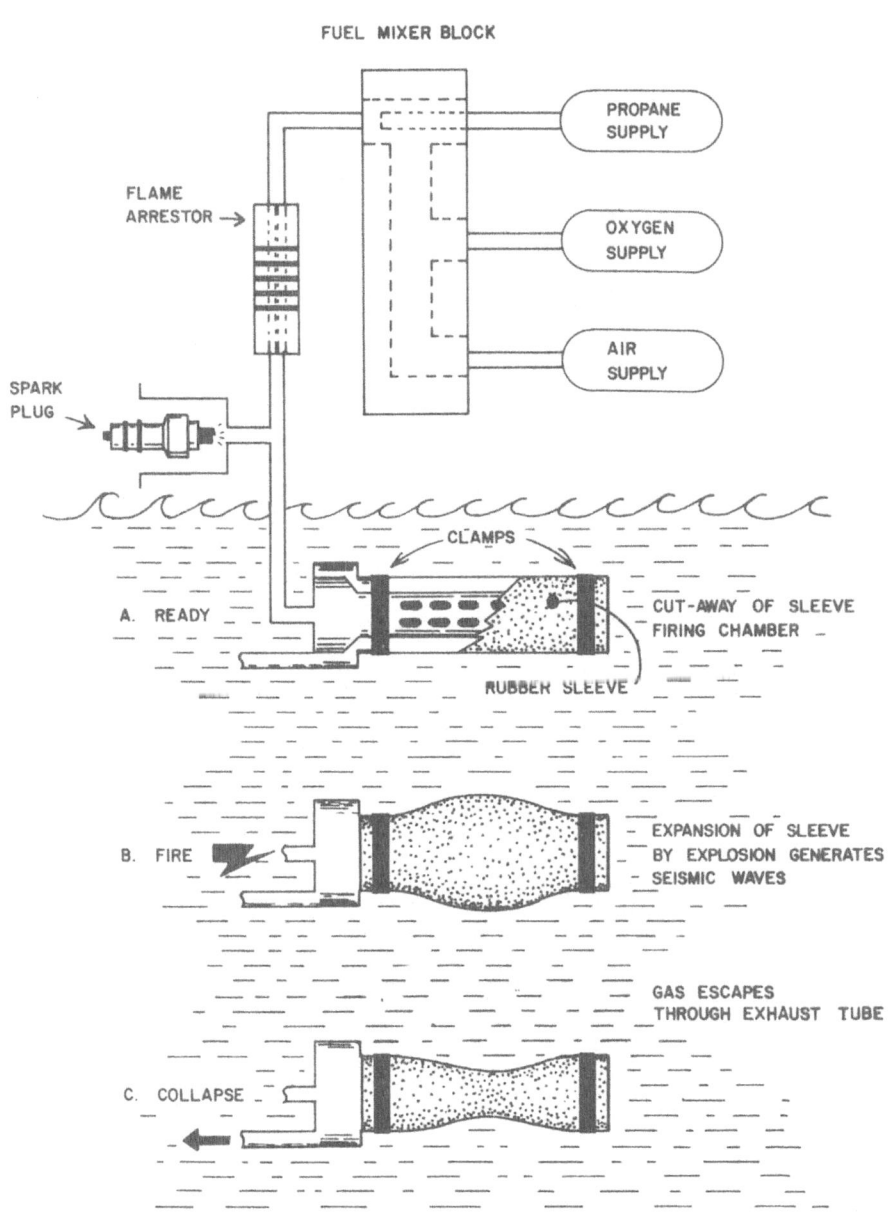

FIGURE 7.12 Schematic illustration (simplified) of Minisleeve Exploder system and operation. (Courtesy of Exxon Production Research Co., Houston, Tex.)

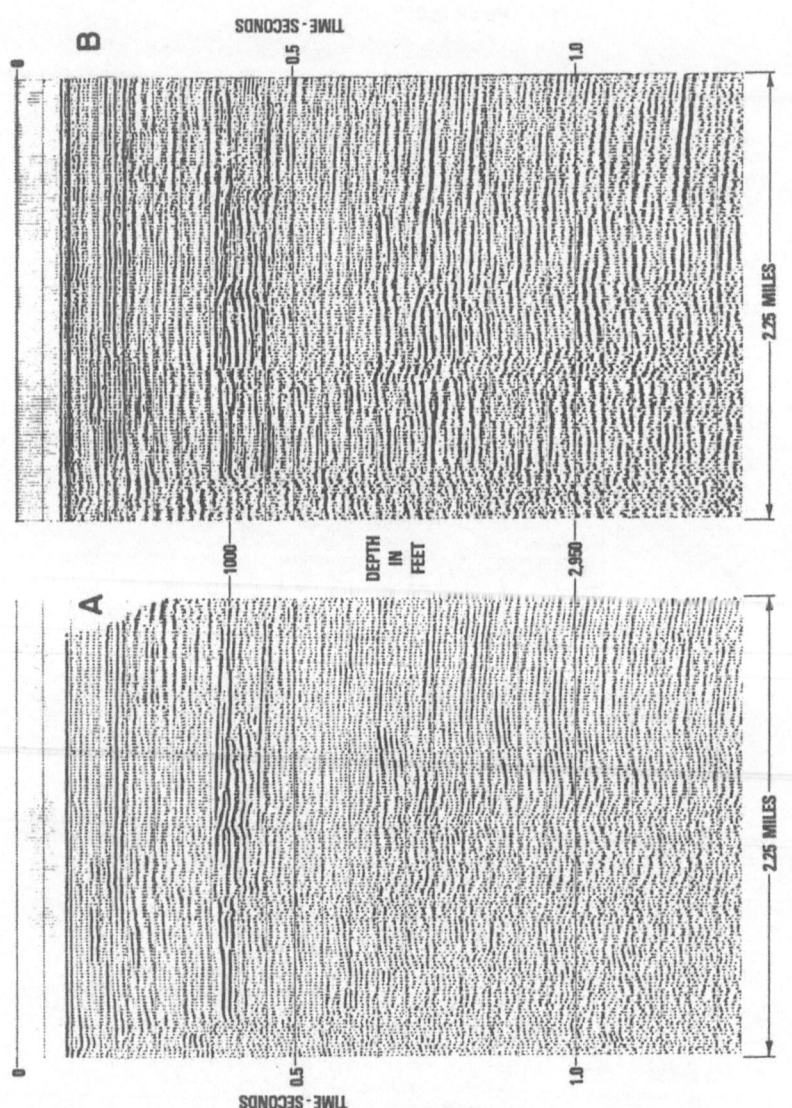

FIGURE 7.13 Comparison of processed multifold HRG profiles, obtained with the Minisleeve Exploder (A) and a 15 kilojoule sparker (B), Gulf of Mexico. (Courtesy of Exxon Production Research [A], and Fairfield Industries, Inc. [B], Houston, Tex.)

GENERAL OPERATIONAL CONSIDERATIONS

The choice of a particular seismic source should involve detailed consideration of the following factors:

Specific area geology and objectives of the survey, such as the maximum desired depth of penetration, and the seismic reflection characteristics of the local sediments.

Logistics of a particular system in terms of weight, ease of deployment, environmental conditions (e.g., water salinity for sparker operation), required support equipment (winches, compressors, compressed gas cylinders, availability of propane and oxygen, etc.).

Data type sought, such as multifold digital or single channel analog, as the signature of some sources may be more amenable to interpretation than others without data processing.

Suitable deployment location aboard the survey vessel, such as the stern or from outriggers. This choice may depend upon the available umbilical connections, and/or the most suitable location with respect to the seismic receiver(s)—hydrophone(s)—and their geometry toward producing quality data.

Recent research has shown that greater signal-to-noise ratios may be obtained by the lateral deployment of seismic sources perpendicularly to the hydrophone array. The use of a survey vessel's wake to break up the multiple returns, by deploying the hydrophone and source at equal distances from midships, is also a frequent practice.

Ease of operation, maintenance, and repair are additional considerations. Above all, the manufacturer's manual should be strictly adhered to for successful operation.

Eight
SINGLE CHANNEL
SEISMIC REFLECTION SYSTEMS

INTRODUCTION

Single channel high-resolution seismic reflection equipment, using the seismic sources described in the previous chapter, have been the principal component for geohazards surveys since their inception. Following World War II, and through the 1950s, industry fostered the development of the continuous seismic profiling technique or CSP (1, 2). In the 1960s further developments of the technique were carried out at major oceanographic institutes.

The technique requires a seismic source (chap. 7), hydrophone, amplification and filtering circuitry, and a method for recording the reflection data. Initially the recorders consisted in rotating drum units upon which a squiggly trace or analog recording was made on paper by a stylus as the drum rotated with time (3). By the early 1960s facsimile recorders were the common method for such recordings, and CSP data were also being recorded on magnetic tape in analog form, using standard high-fidelity reel-to-reel recorders.

The following discussion of the equipment and operation of single channel systems is subdivided into sections on the hydrophone receiver elements, analog processing (signal enhancement), recording methods, operational considerations, and interpretation pitfalls.

HYDROPHONE RECEIVERS

Acoustic signals transmitted by a surface towed seismic source, and reflected from the sea floor and underlying strata, are detected on the surface, or at depth in the case of deep-towed systems, by a hydrophone. The hydrophone consists of an array of piezo-electric acoustic crystals. The crystals are pressure sensitive and generate a small electric current in response to the pressure of the returned acoustic signals and ambient noise. Hydrophone cables contain a number of such crystals, attached in parallel, in order to increase signal reception, while reducing interference from ambient noise.

The piezo-electric crystals used in hydrophones should be sensitive to a wide frequency spectrum, and have a broad flat band response (equal at all frequencies) for signals between 30 and 3000 Hz. Of primary concern in the design of hydrophone arrays is the signal-to-noise ratio, and elimination or reduction of as much as possible of the ambient noise created by the towing of the receivers through the water. Thus, the crystals are usually packaged within a flexible plastic membrane (fig. 8.1) filled with a nonelectrolitic fluid, such as kerosene or jet aircraft fuel (JP). This fluid prevents electrical signal leakage between the wires, and

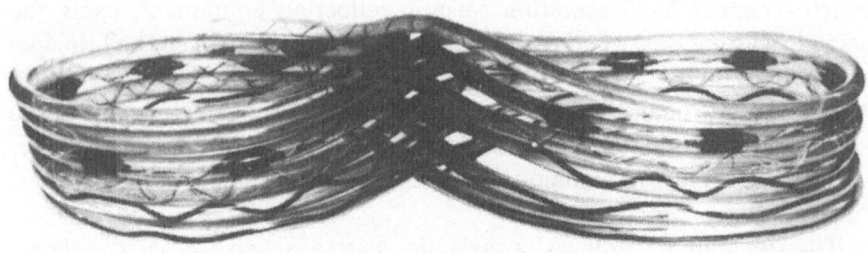

FIGURE 8.1 Single channel hydrophone streamer: 7/8 inch (2.22 cm) diameter polyvinyl-chloride tubing, encasing 20 piezo-electric lead zirconate/titanate transducers, spaced 6 inches (15.24 cm) apart, with 20 dB preamplifier. Frequency response of the streamer is 50–3000 Hz, and the sensitivity is –103 dB referenced to 1 microbar/volt at one yard. (Courtesy of Innerspace Technology, Inc., Waldwick, N. J.)

provides the necessary density to balance the cable in the water, while providing optimum acoustic coupling with the water column. The hydrophone cable is towed sufficiently aft of the survey vessel to reduce the effects of ship-generated noises, such as those created by the propellers and vibration of motors.

Amplification of the recorded signal within the hydrophone streamer is done by a wide-band preamplifier. Solid-state units are both compact and reliable, consisting of a single chip or integrated circuit, and a small battery. While preamplifiers have been strictly analog to date, the use of digitizing circuitry within the hydrophone cable and transmission of the signal in digital bits to the survey vessel is becoming common practice.

The hydrophone cable needs to be set below the sea surface in order to reduce noise caused by surface waves and inherent cavitation, and increase acoustic coupling as a result of greater hydrostatic pressure. The buoyancy of the hydrophone cable must also be balanced in order for it to be maintained at a selected depth. The weak seismic reflection signals, produced at the hydrophone crystals, are further amplified and filtered aboard ship to increase the signal-to-noise ratio before being recorded.

ANALOG SIGNAL PROCESSING

Optimum tuning of the signal prior to recording requires both experience and trial and error adjustments of the filters and amplitude gain settings for specific seismic sources and survey areas. Thus a low-cut (reduction of signals below a certain frequency) filter setting of 300 Hz and hi-cut (reduction of signals above a certain frequency) of 800 Hz may provide optimum signals for data recording in one geographical area for a particular seismic source, while proving inadequate in another area or for a different seismic source.

In addition to filtering and overall gain settings, the seismic reflection signals tend to decay as a function of time, due to spherical spreading of the acoustic waves away from both the source (outgoing pulse) and reflecting horizons, and to absorption of the signal within the sedimentary strata. This decay is theoretically equivalent to the inverse of the cube of the signal travel distance, and must be corrected for. Such amplification or gain compensation is applied in practice with respect to time rather than distance. The gain circuit that compensates for such acoustical signal attentuation consists of a gain ramp or slope whereby the gain is boosted with respect to record time, and is referred to as TVG or time varying gain. Or in some cases, it is applied automatically by the amplifier circuit known as AGC or automatic gain control.

Such circuitry and their adjustments thus provide a continuous increase in gain with respect to time or depth on a seismic reflection record, boosting the the signal strength of deeper reflectors to values close to those near the sea floor. The effect of an AGC circuit is illustrated by the seismic profiles of figure 8.2 (A and B). Additional signal adjustments are performed at the recorder to assure a clear display of incoming subbottom seismic reflection signals.

ANALOG RECORDERS

As stated earlier, facsimile type graphic recorders replaced the wiggly line single trace drum systems for recording analog seismic signals. A number of recording instruments are available (fig. 8.3), and consist in a mechanically rotating stylus belt or helix that moves across the recording paper (either wet or dry), sandwiched between the stylus and a ground bar, which acts as a ground to conduct the amplified signal from the stylus through the paper. The recorder thus produces a cross-sectional display of reflection time (stylus sweep) versus distance along the survey track (paper advance/survey vessel speed).

Modern recorders allow for a multitude of selections and adjustments of stylus sweep rates and trigger outputs for firing the seismic source. Some models include time delay circuitry, which allow the expansion of scales to provide optimum display of the desired portions of seismic reflection profiles. The addition of swell filtering devices (chap. 6) to the recorder input helps produce clearer recordings when surveying in rough seas.

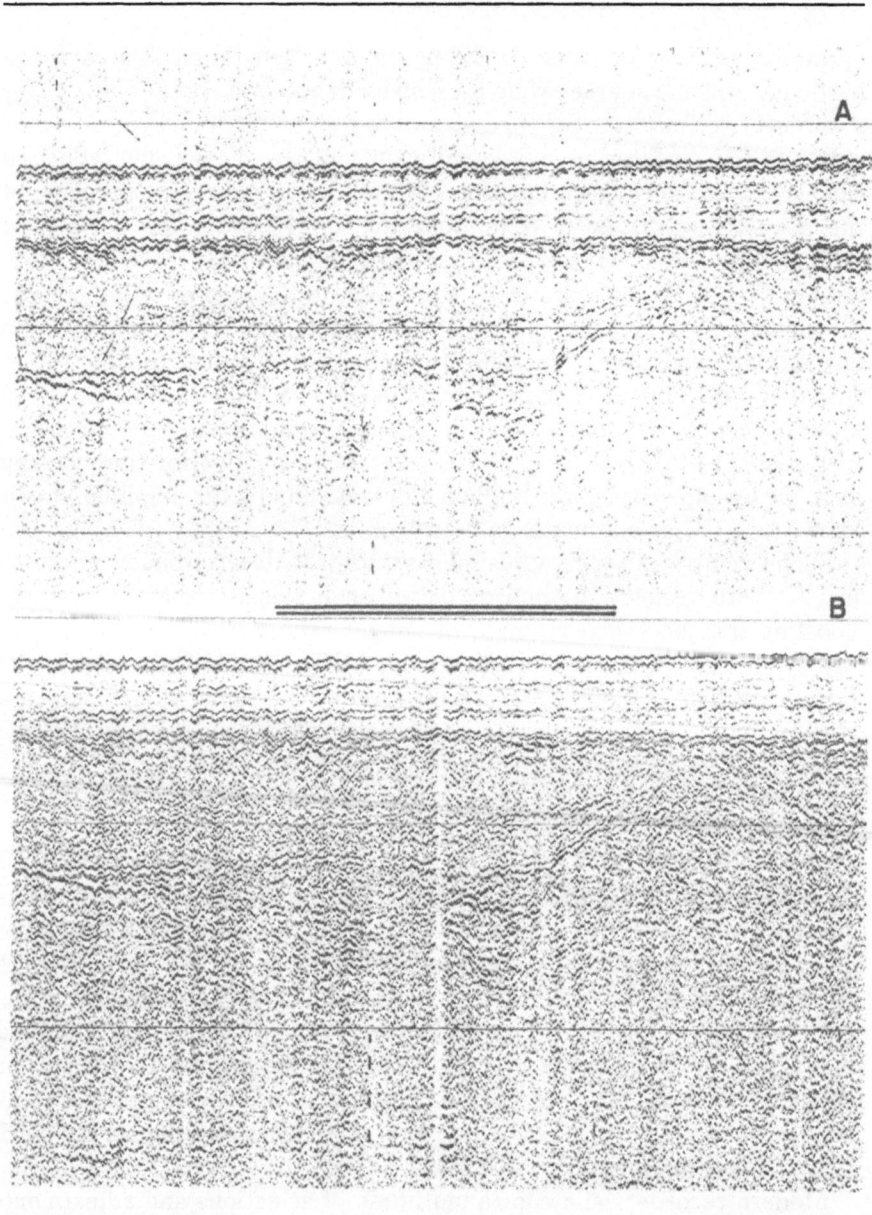

FIGURE 8.2 Comparison of HRG analog seismic sections showing effect of: (A) uncompensated gain, and (H) compensated gain amplification, for 300-joule Uniboom (tm-EG&G) source, North Sea. (Courtesy of Technical Survey Services, Ltd., Croughton, Northamptonshire, U.K.)

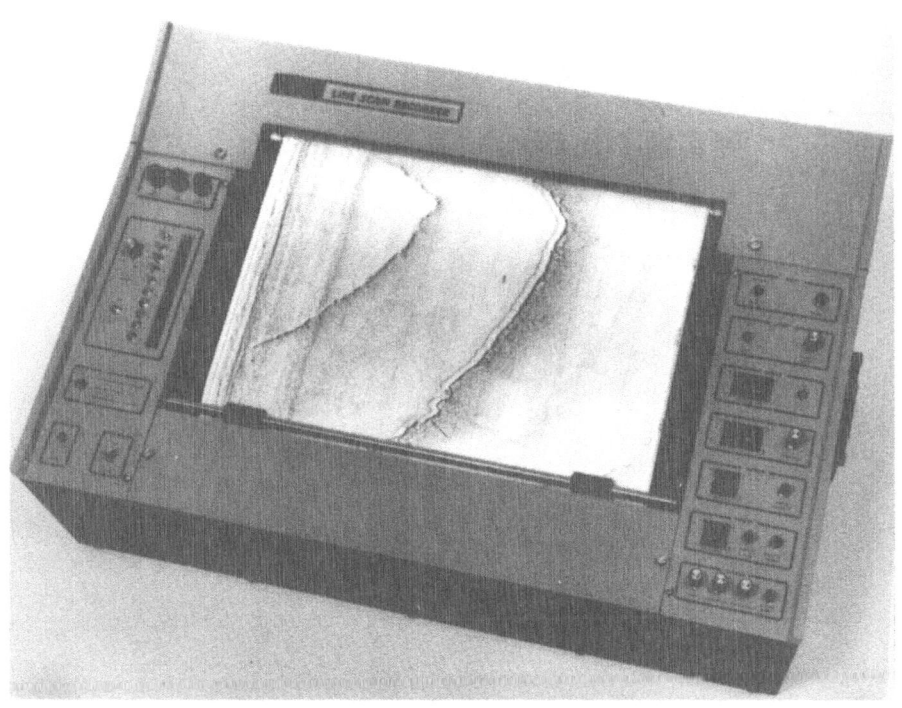

FIGURE 8.3 Facsimile-type recorder used for analog HRG seismic reflection data display. (Courtesy of Raytheon Company, Portsmouth, R. I.)

The recorded scale needs to be annotated, in order to determine the vertical exaggeration (ratio of true vertical to horizontal scales), as well as the vertical scale in terms of time units (e.g., 100 msec divisions, or 0.5 sec recording). Other pertinent information including filter and gain settings, delay times, location, time and date, and survey line number should also be noted on the field records as well as any changes in these settings during the course of a particular survey line.

While the simplicity of such recording devices attests to their popularity, the printing density or dynamic range is limited to about 24 dB, which is a limiting factor of the analog recording technique (2). Recent improvements in the ability of recording systems to accommodate larger dynamic ranges with the use of digital processors are, however, quickly bridging this gap.

More recent developments include the production of fiber optic and thermal type printer/recorders, while the thrust of electronic development

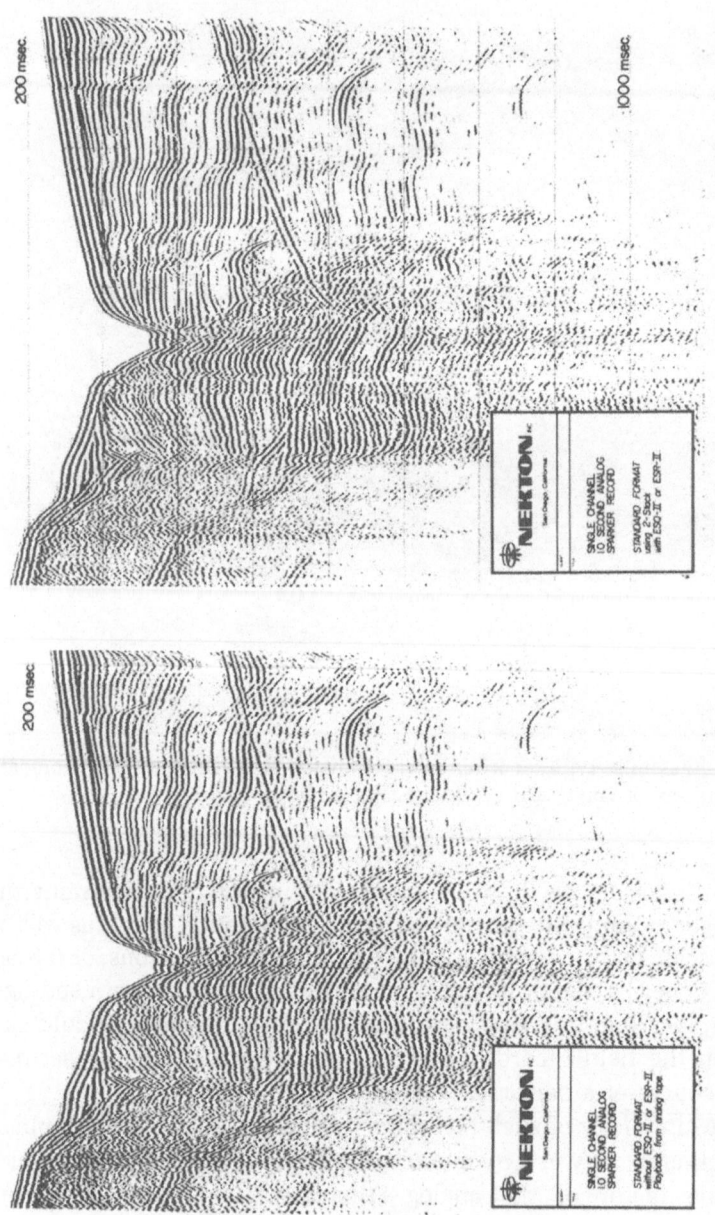

FIGURE 8.4 HRG single channel sparker seismic profile, recorded: (A). in standard format, and (B), after real-time processing (two-trace stack, or summation), offshore California. (Courtesy of Nekton, Inc., San Diego, Calif.)

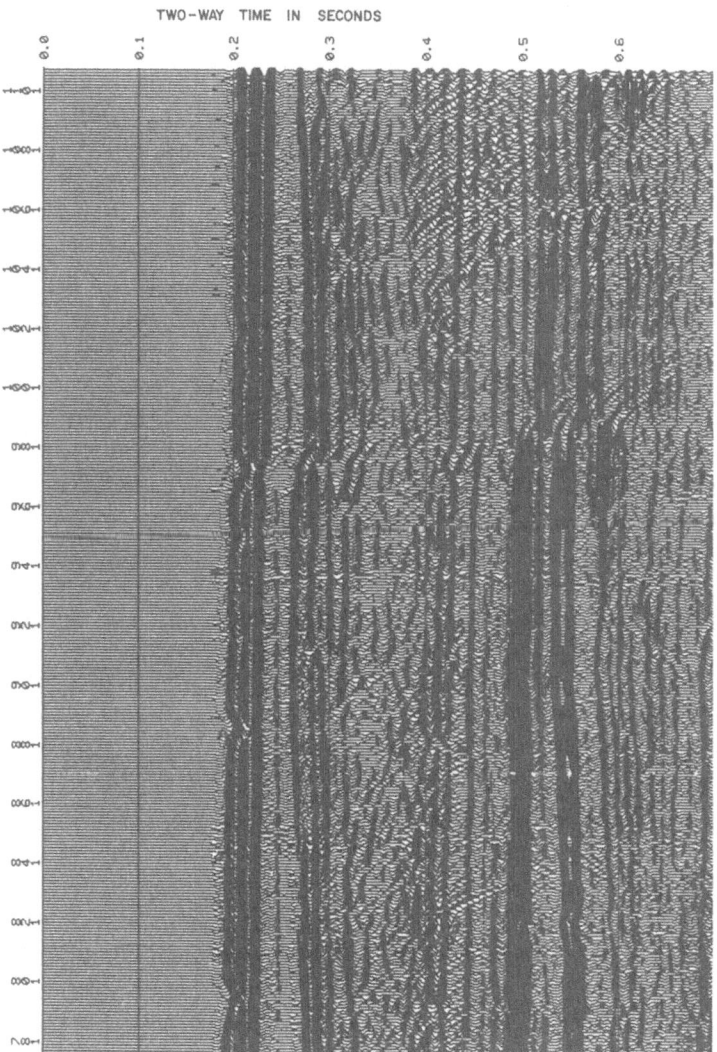

FIGURE 8.5 Multichannel HRG profile processed in real time during survey acquisition, offshore Venezuela. (Courtesy of BEICIP/CGG/INTEVEP.)

will produce digital systems with capabilities far superior to the present popular single channel analog systems.

REAL-TIME DIGITAL SIGNAL PROCESSING

Present signal processing is generally limited to gain controls and one-time filtering on prerecorded data, and some vertical stacking or smearing of multiple shots (fig. 8.4). Microprocessor-based signal processing systems that greatly enhance the acquisition of CSP data, however, are limited in use. The computer processing sequences now performed ashore on multifold seismic reflection data can, to a certain limit, be performed in *real time* aboard ship during a survey (fig. 8.5). Optimum filter selection, which may vary with time or depth, true amplitude gain recovery, and deconvolution of received seismic signals to produce a sharp spike (compressed pulse), may all become real-time processes in the near future.

OPERATIONAL CONSIDERATIONS

Optimum deployment and proper adjustment of recording parameters (AGC, gain, threshold and filter settings) of a single channel seismic profiling system permit a high signal-to-noise ratio at the final analog recording, with good resolution of the stratification beneath the sea floor.

The spacing and arrangement of the source and hydrophone are critical for maximum data quality. If these are too far apart the first arrival or direct water wave from the source may mask concurrent arrivals of reflections.

The depth at which both the source and receiver are towed is critical in eliminating bubble pulses and ghosts (secondary signals reflected from the sea surface). Reflected signals may also be filtered out or masked, depending upon the submerged depth of the hydrophone cable, since reflected signals are recorded when they arrive from the sea floor as well as from the sea surface after passing the hydrophone. This effect, termed *aliasing*, may be eliminated by towing the hydrophone at a depth equivalent to one quarter of the wavelength of the desired frequency (fig. 8.6). Thus, the cable depth should be equivalent to a quarter wavelength of the desired reception frequency. This depth may be computed using an appropriate speed of sound in water, such as 1500 msec, and computing the wavelength for the desired frequency by the equation:

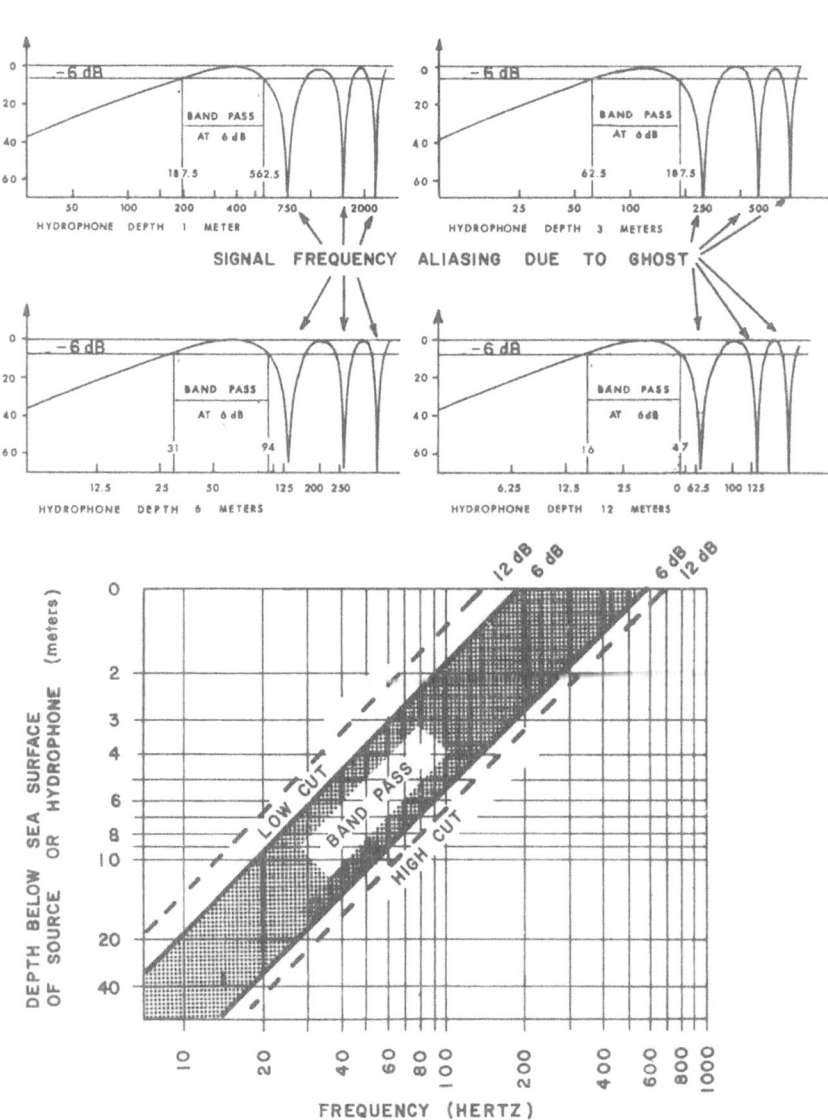

FIGURE 8.6 Graphs (*upper*) depicting effect of frequency aliasing (filtering) caused by ghosting of signals between sea surface and source or hydrophone (at depths of 1, 3, 6, and 12 meters). Graph (*lower*) depicting frequency band pass as function of hydrophone and/or source depth below sea surface. (Courtesy of BEICIP/RUEIL/MALMAISON.)

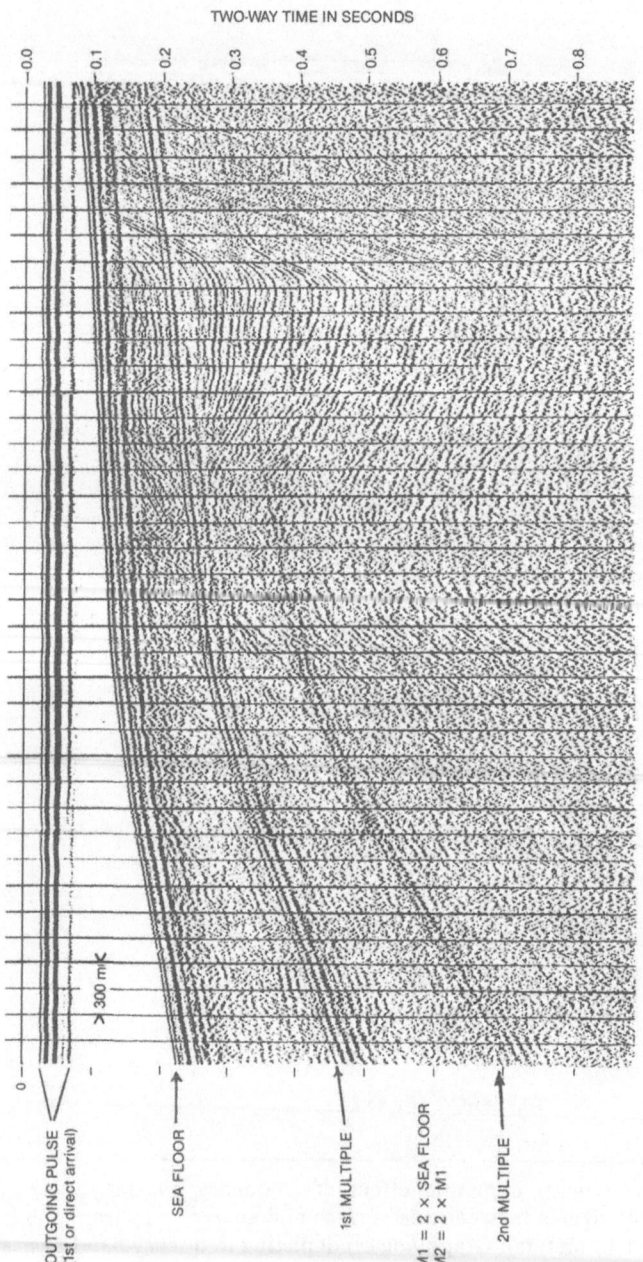

TWO-WAY TIME IN SECONDS

0.0 0.1 0.2 0.3 0.4 0.5 0.6 0.7 0.8

> 300 m

OUTGOING PULSE
(1st or direct arrival)

SEA FLOOR

1st MULTIPLE

M1 = 2 × SEA FLOOR
M2 = 2 × M1

2nd MULTIPLE

FIGURE 8.7 Single channel analog record, showing multiples and their interference. Minisleeve exploder source (Exxon Pat.), Point Conception area, offshore California. (Courtesy of COMAP Geosurveys, Inc., Houston, Tex.)

$$\lambda = 4 \ (f/v),$$

where λ is a quarter of the signal wavelength, f the center or average frequency, and v the velocity of sound. If the frequency is 500 Hz, a quarter wavelength would be 0.75 m, at which depth the hydrophone would enhance incoming signals at this particular frequency. By the same token, a cable depth equivalent to one-half wavelength would tend to cancel signals for a given frequency (fig. 8.6).

The advantage of having a number of hydrophone receivers strung out over a distance (few meters), is that they will receive signals from the sea floor simultaneously, while signals arriving from other than normal angles will arrive with small offsets in time, which if properly spaced (1/4 wavelength) will cause mutual interference and elimination. If the spacing between individual phones is variable, this technique allows the cancellation of noise from a greater segment of the frequency spectrum and helps produce higher signal-to-noise ratios.

The dielectric fluid (usually kerosene) used to fill the hydrophone cable should produce a neutral buoyancy relative to the surrounding water (salt or fresh) to maintain the cable at a fixed depth during a survey. The polarity of the electrical connections between the individual phones within a streamer must also be correct (in parallel) or the incoming signals will cancel out.

The use of inactive cable sections, which are either dummy or vacant unused sections, both ahead and aft of the active hydrophones, provide a streamlining effect and act as shock absorbers to dampen surges and vibrations (strumming) of the survey vessel motion.

INTERPRETATION

While appearing relatively straightforward, the interpretation of single channel analog records requires a thorough understanding of both the physics of the seismic reflection profiling technique as well as the mechanics and electronics of the recording system. Errors or pitfalls common to such interpretations involve the mistaken identification of multiple reflections as well as seismic source pulse oscillations for subbottom stratification (fig. 8.7).

The true scales of the geological and bathymetric features are easily overlooked as a result of vertical exaggeration of the recordings; thus, relatively low dips may appear as catastrophic precipices on CSP records. The true depths to features may be determined only if velocity data are

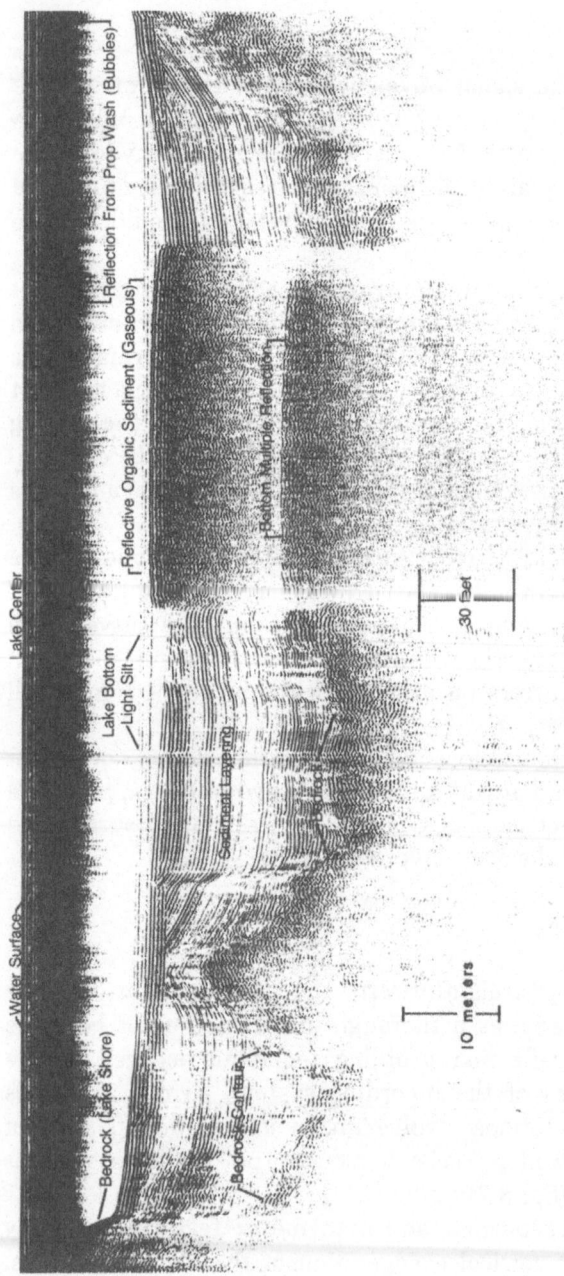

FIGURE 8.8 Single channel analog recording, made with compact sparker (25 joule) system, showing effect of gas-bearing sediments, shallow stratigraphy and bedrock, Greenwood Lake, New Jersey. (Courtesy of Innerspace Technology, Inc., Waldwick, N. J.)

available, while the use of estimated velocity profiles (typically 1500 m/sec) can only yield an approximation of true depth.

The presence of gas within near-surface sediments may also be misinterpreted as stratification and/or structure, while merely indicating the upward limits of gas migration, as shown in the profile of figure 8.8. Diffraction patterns from pipelines or boulders, and reflections from relief features adjacent to the survey line, must not be confused as true seismic reflections.

A combination of experience and understanding of basic geophysical principles coupled with the integration of engineering borehole data are an invaluable aid toward the understanding of CSP profile records.

Nine
MULTIFOLD ACQUISITION
AND DIGITAL PROCESSING

The use of multifold digital acquisition systems for high-resolution seismic reflection surveys became standard practice during the 1970s. The technique was developed in the early 1960s as an appreciable improvement over the single channel continuous seismic profiling method. The precedure and technique are similar to the acquisition of deep exploration data (thousands of meters) for the petroleum industry.

The digital sampling of HRG seismic reflection data, however, is on the order of a msec or less, and total record lengths are only one to two sec, compared to 4 msec and 6 to 8 sec for exploration data. The hydrophone cables and group intervals employed are also shorter, as well as physically smaller (3 to 4 cm, o.d.) compared to those systems used for petroleum exploration that are 6 to 9 cm, o.d.

Following is an explanation of the multifold hydrophone system, the field acquisition procedure, and digital processing of multifold seismic reflection data.

MULTIFOLD SEISMIC RECEIVER ARRAYS

Hydrophone Cables
Hydrophone receivers used for multifold data acquisition consist of arrays from 6 to 48 groups, each carrying out the function of the single channel hydrophone discussed in the previous chapter. Within the multifold array, streamer or cable, the individual groups are hardwired to the digital acquisition system recorder aboard ship. As the number of groups is increased, the cable and electrical connections become proportionately larger to accommodate the additional conductors.

As with single trace systems, the basic requirement for seismic hydrophone receiver arrays, employed in the acquisition of CSP data at sea, is that they be sensitive to signals within the desired frequency spectrum, while still maintaining a high signal-to-noise ratio. The curve of frequency versus pressure sensitivity for such hydrophones should display a flat portion of equal receptivity within the frequency range of 50 to 1000 Hz., as shown in figure 9.1.

Hydrophone Group Geometry
Multifold hydrophone cables consist of a number of sections each containing several groups of individual pressure sensitive (piezo-electric) hydrophone elements. The arrangement and number of hydrophones may vary. Individual groups or phones within a cable are wired in series to

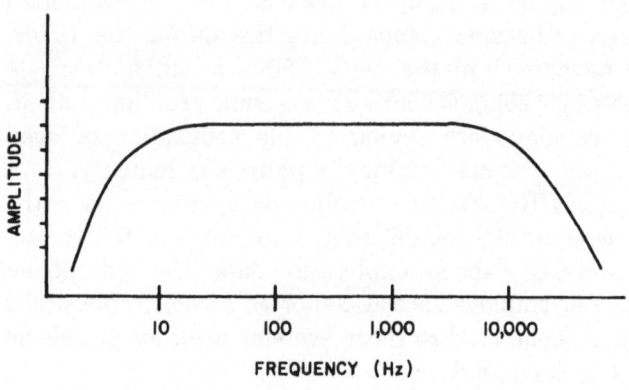

FIGURE 9.1 Frequency response curve for typical HRG hydrophone.

produce a receiver array of a particular length, which constitutes a single trace equivalent to a single channel hydrophone. The hydrophone elements, associated signal conductor wires, bulkheads (for routing stress members and maintaining shape at 30 to 100 cm intervals), and stress cables are set within flexible plastic tubing filled with an electrically nonconductive fluid such as kerosene or jet aircraft fuel.

The number of individual receiver groups has increased over the past decade, with 12 and 24 groups being common at present compared to only six groups at the end of the 1970s. The spacing of individual groups along the hydrophone cable streamer is adjusted to meet the required objectives of a particular survey, such as depth of penetration, distance between successive shots (firing interval), and water depth (to remove the multiples).

For example, if the desired subsurface coverage is on the order of 0.5 sec, and within shallow waters of 100 m or less, a short group spacing would be desired, whereas for deeper waters and subsurface objectives of 1 to 2 sec, a longer group interval spacing would be required.

The need for such adjustments in receiver geometry is to provide an appropriate offset (NMO) in the arrival times of multiples and other unwanted signals, so that these may be removed during the stacking of CDPs in data processing. If the spread or distance between near and far traces (closest and farthest from the energy source), is too small during data acquisition to provide sufficient offsets in the arrival times of multiples, these will be difficult to remove. This results in reducing the effectiveness of multifold processing for attaining significant improvements in the signal-to-noise ratios and the elimination of undesirable reflection signals such as multiples.

Hydrophone Streamer Depth

Besides the hydrophone array geometry (separation and location of individual groups), critical attention must be placed upon the depth of the hydrophone streamer. Depth-controlling birds or ballast (for less noise) are attached to the hydrophone streamer in order to maintain a constant depth of submersion along the entire cable length, according to either preset values or by remote control from the recording laboratory. Calibrated depth transducers or depth indicators (DIs) permit monitoring the streamer depth at selected locations along the cable. For a 12 or 24 group HRG cable, typically 300 to 500 m in length, the birds and DIs are commonly located at the front, center, and back of the active portions of the cable.

A critical consideration in the selection of cable depth is the filtering or aliasing effect produced by ghosts. This effect is due to the reflection of seismic signals off the sea surface, after passing the submerged hydrophone cable, and their recording as a delayed or ghost signal as discussed in the previous chapter (fig. 8.6). The distance between the cable and the sea surface is thus critical. Primary reflectors attain a multiple leg appearance, which is dependent upon the depth of both the source and receivers, due to the ghost signals.

Optimum streamer depth is approximately equal to a quarter wavelength for the desired signal frequency (fig. 8.6). Within the broad frequency bandwidth of HRG seismic recordings, this becomes impossible when all hydrophones are towed at a constant depth, as certain frequencies will be enhanced and others cancelled (fig. 8.6). For example, a streamer depth of 3.75 m would enhance incoming signals at a frequency of 100 Hz, and other harmonics of this frequency. By the same principle, certain frequencies are invariably filtered out or aliased during field acquisition (fig. 8.6).

However, if the hydrophone streamer cable is towed in the water column at an angle to the sea surface behind the survey vessel, the aliasing effect can be alleviated. This may be done by using the FLAIR (patented by Fairfield Industries, Inc.) technique, whereby the near trace group is submerged at a shallower depth than the far groups. Through the application of appropriate static (depth of cable) corrections to the respective groups during data processing, phase shifts are used to cancel the ghost signals. This process permits the reception of frequencies along a more normalized curve, where reception is practically flat from 40 to 250 Hz, and the previously cut-out portions of the frequency spectrum have been eliminated.

Experimentation along these lines may bring forth further developments within the high-resolution field of nonlinear hydrophone spacing of cable groups to further enhance signal-to-noise ratios.

The reduction of streamer friction and resultant noise by the application of nonstick, or water absorbent coatings, may further enhance present signal-to-noise ratios, and allow greater survey vessel speeds during HRG data acquisition.

Digital Hydrophone Cables

It should be expected that by the late 1980s the use of analog-to-digital conversion circuitry, within the cable, and transmission of the multiplexed digital data, by means of fiber optic cables, will become common practice. These techniques allow the transmission of all the data from individual hydrophones on a single pair of lightweight fibers, instead of a large bundle of conductor wires, enabling the use of more phone groups while easing both deployment and maintenance (1).

DIGITAL DATA ACQUISITION AND STORAGE

Incoming analog reflection signals (and noise) to the hydrophone cable arrive at the instrument shed aboard the survey vessel within a control panel (fig. 9.2), allowing the testing of the individual phone groups to ensure proper operation. The use of an oscilloscope also allows the continuous monitoring of all the phone groups during the course of a survey. The data then proceed through an analog-to-digital converter and editor prior to being recorded on magnetic tape (fig. 9.2).

The details of these processes are covered in the literature (2), and by the manufacturer's operating and maintenance manuals, along with the correct and necessary testing procedures that should be performed at regular intervals to ensure that the data are properly recorded, and that the electronics are performing to par with preestablished specifications.

One of the most common tests for assuring proper data recording is the oscilloscope camera monitor record (fig. 9.3), which displays the incoming signals from all channels for a single shot. This display permits an immediate assessment of the recording operation. Unwanted signals such as 60 (or 50) cycle interference (caused by electric leakage), inoperative phone groups, and reversed polarity wiring, are immediately noticeable to

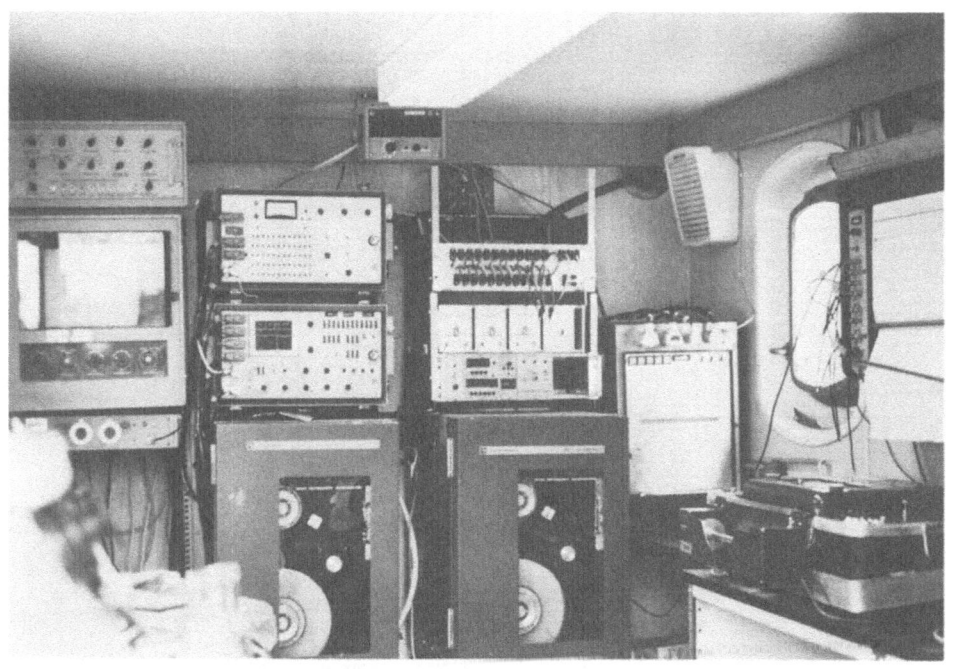

FIGURE 9.2 Shipboard multifold digital recording instruments, and hydrophone input/control panel. (Courtesy of COMAP Geosurveys, Inc., Houston, Tex.)

the trained eye. Detailed checks of the equipment and recording instruments during a survey are critical since if the data are not recorded properly, they may be of no use later when the time comes for shore-based processing.

Digital Sampling
The process by which incoming analog seismic signals are sampled and assigned binary numbers or digits is referred to as *digitizing*. The process involves an assessment of the amplitude of a signal (in millivolts) with respect to time (e.g., quarter millisecond), and the output of digital numbers for storage on magnetic tape, and processing within a digital computer system (2).

Digital sampling of the seismic reflection waveforms should allow the full restoration of the original reflection signal. Sampling should ideally be

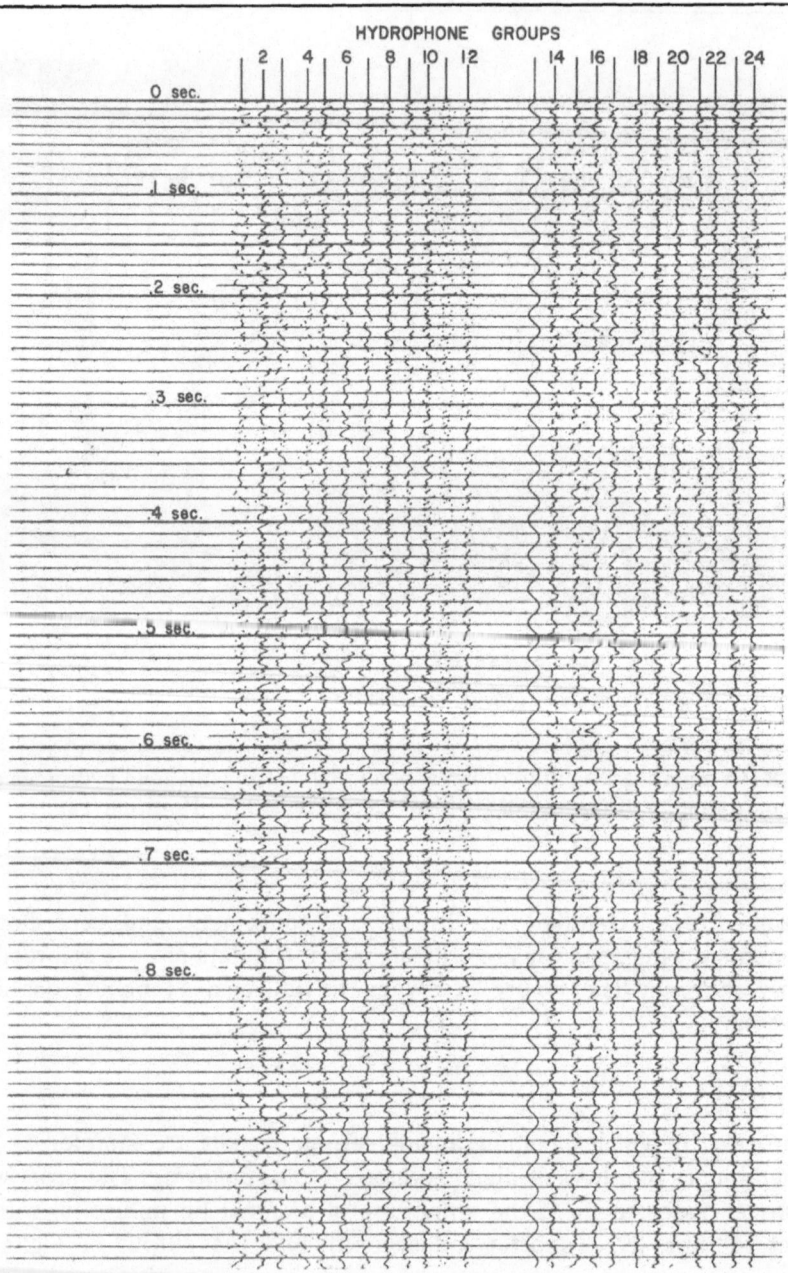

FIGURE 9.3 Camera monitor record, for 24 trace multifold seismic system. Note noise on all traces and 60 Hz interference on group 13. (Courtesy of COMAP Geosurveys, Inc., Houston, Tex.)

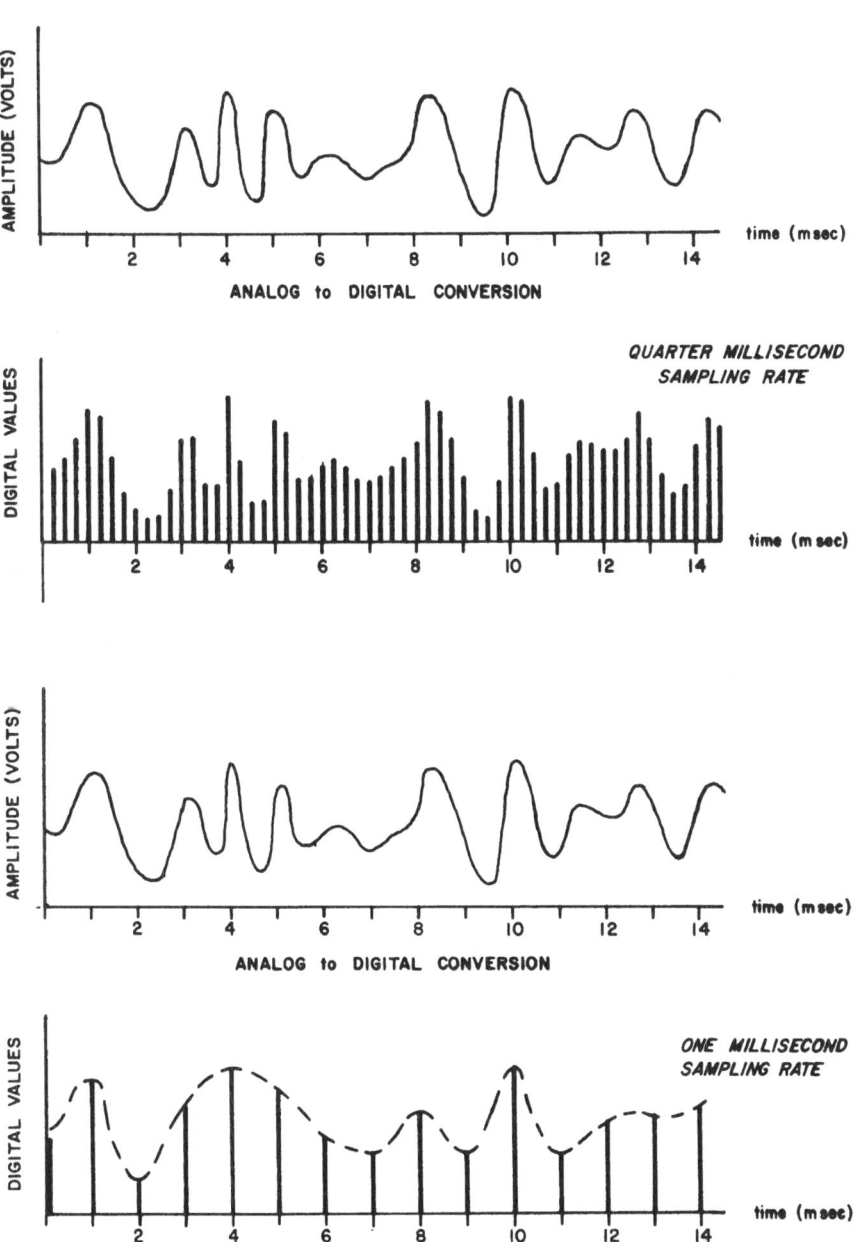

FIGURE 9.4 Digital sampling of analog data. Quarter-mil sampling (*top*) defines signal without loss of fidelity. One-mil sampling (*bottom*) results in incomplete digital sampling, aliasing the higher frequencies.

on the order of a quarter wavelength or less (see fig. 9-4 top). If fewer samples are taken, aliasing (data filtering and deterioration) occurs, and fidelity of the original signal is lost (fig. 9.4 bottom). Thus, a digital sampling rate of 1.0 msec will exclude signals of frequencies greater than 500 Hz and produce a low-fidelity recording. This relationship is illustrated in figure 9.4, where lower sampling rates for a given frequency will result in aliasing. The standard sampling rate for seismic exploration data, for example, is on the order of 4 msec, while that for HRG data is 0.5 or 1.0 msec. It is also becoming common to sample HRG seismic data at rates of up to 0.25 msec (quarter mil data) from a near trace or separate hydrophone group, in order to preserve a very high-resolution seismic reflection picture of the immediate subbottom to about 200 m below the sea floor.

Another benefit in the recording of digital high-resolution seismic data is the retention of true amplitude. To date, single channel analog recordings have essentially been limited in this respect. The maximum amplitude gain offered by analog recordings, as pointed out in the previous chapter, has been limited to approximately 20 to 25 dB. Digital seismic recording systems of the late 1970s, operating on a fixed gain technique, allow for as much as 80 dB of total gain retention, whereas modern floating point gain systems allow for a full 100 dB amplitude retention through the digital recording process. This latter value is about the limit in terms of necessary amplitude recovery requirements for the sampling of HRG seismic data.

The retention of large variations in signal gain is particularly useful for the generation of relative or preserved amplitude (RAP) seismic sections. Gas-bearing strata or bright spots are frequently identifiable on such processed records. The RAP processing has provided a powerful tool in the seismic exploration for hydrocarbons over the past decade (3), as well as in HRG hazard surveys.

Standard recording instruments, employed in the acquisition of exploration seismic field data, are generally suitable to recording higher frequencies and sampling rates associated with HRG data. The shorter sampling intervals used in the digital recording of HRG data are needed to retain the higher frequencies with the appropriate fidelity for reproduction (fig. 9.4).

Present-day digital field recording systems (fig. 9.2) provide the best of the above parameters, as well as the versatility to be modified during the course of a survey to suit the particular geological environment and geoengineering survey objectives.

PROCESSING OF MULTIFOLD DIGITAL SEISMIC DATA

The processing of HRG multifold seismic reflection data in the early 1980s is practically identical to that employed for exploration seismic data. The only difference between the two involves the smaller digital sampling rates of from 0.25 to 1.0 msec for a 2.0 sec record, compared to 2.0 to 4.0 msec, and a 6.0 to 8.0 sec record for exploration seismic data. In some cases an effective reduction in processing costs may be obtained by resampling the 1.0 msec field data at 2.0 msec without a significant loss in resolution.

Data Processing

The processing of multifold HRG seismic reflection data usually follows a sequence of operations including most, but not necessarily all, of those discussed in the following. In order to illustrate this process, an HRG multifold (24 channels, miniseleve exploder seismic source, recorded to 2.0 sec) seismic profile (Line 010) from offshore Santa Barbara, California, is processed and displayed for each processing step.

1. *Demultiplexing* of original field tapes consists in the initial sorting of digital field data into individual reflection traces (fig. 9.5). The data were originally split up (multiplexed) during field recording, in order to fit a continuous flow of data onto tape from a number (24 in this case) of individual recording hydrophone groups, following the analog to digital conversion at 1.0 msec intervals. Each of the eight panels of fig. 9.5 represent a single 24 channel recording, as recorded in the field. Note that channel 14 has been killed (removed) because the channel was dead (absent) on the field recordings. The first arrival can be seen at the top of each shot, as well as noisy traces in the first, second and fourth panels from the right.

2. Figure 9.6 is a display of the *near trace* playback for line 010, just as the record would appear on the near trace monitor record aboard ship. This is essentially a single channel display of the data and represents the starting point for data enhancement. Subsequent processing steps should be designed to clean up and improve the quality of this record (fig. 9.6).

3. *Autocorrelation* of near traces provides a graphic display (fig. 9.7) of the similarity of the recorded signals for the entire line (with respect to a zero time for maximum correlation). It is based upon a mathematical function (2), and serves to select an appropriate window length (in msec) for establishing a deconvolution operator,

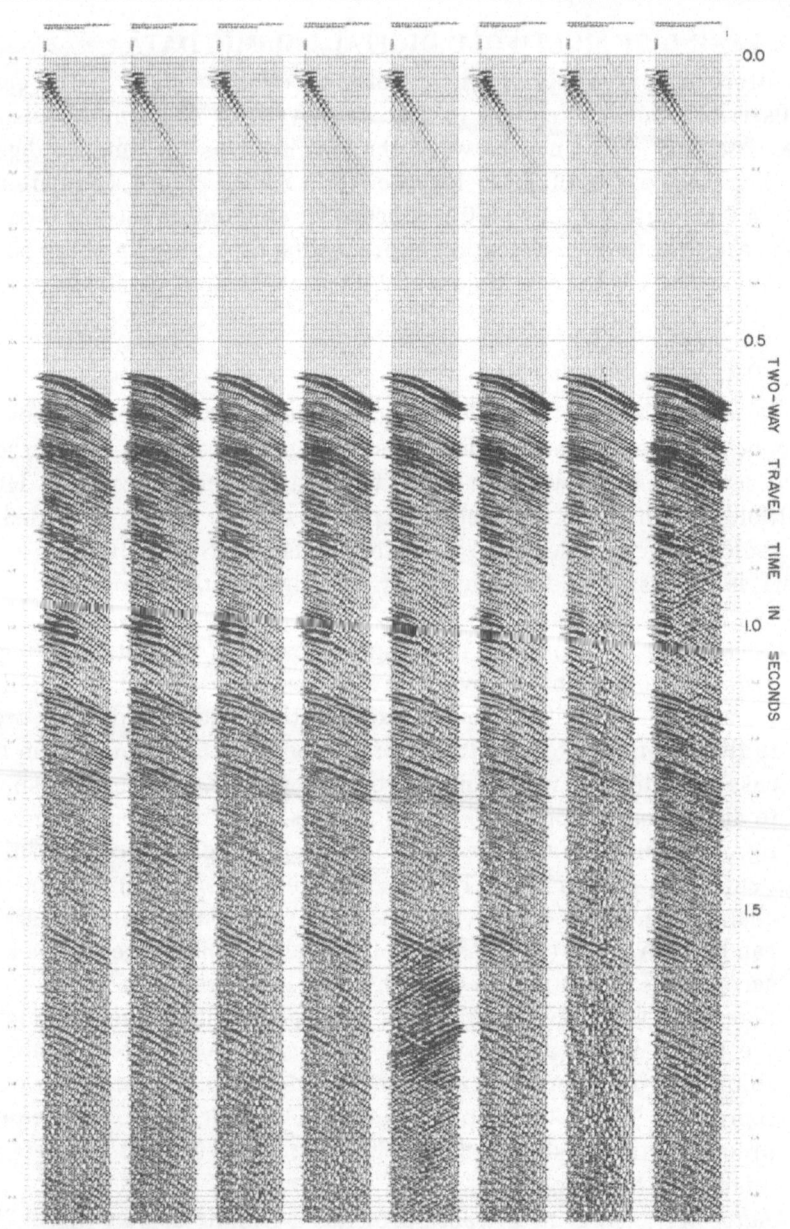

FIGURE 9.5 Display of demultiplexed data for 24-channel record at eight consecutive shotpoints. Record length 2048 msec with 1-msec sample interval recover ramp of: T^1. Line 010, offshore Santa Barbara, California. (Courtesy of COMAP Geosurveys, Inc., Houston, Tex., processed by Sytech Corp., Houston, Tex.)

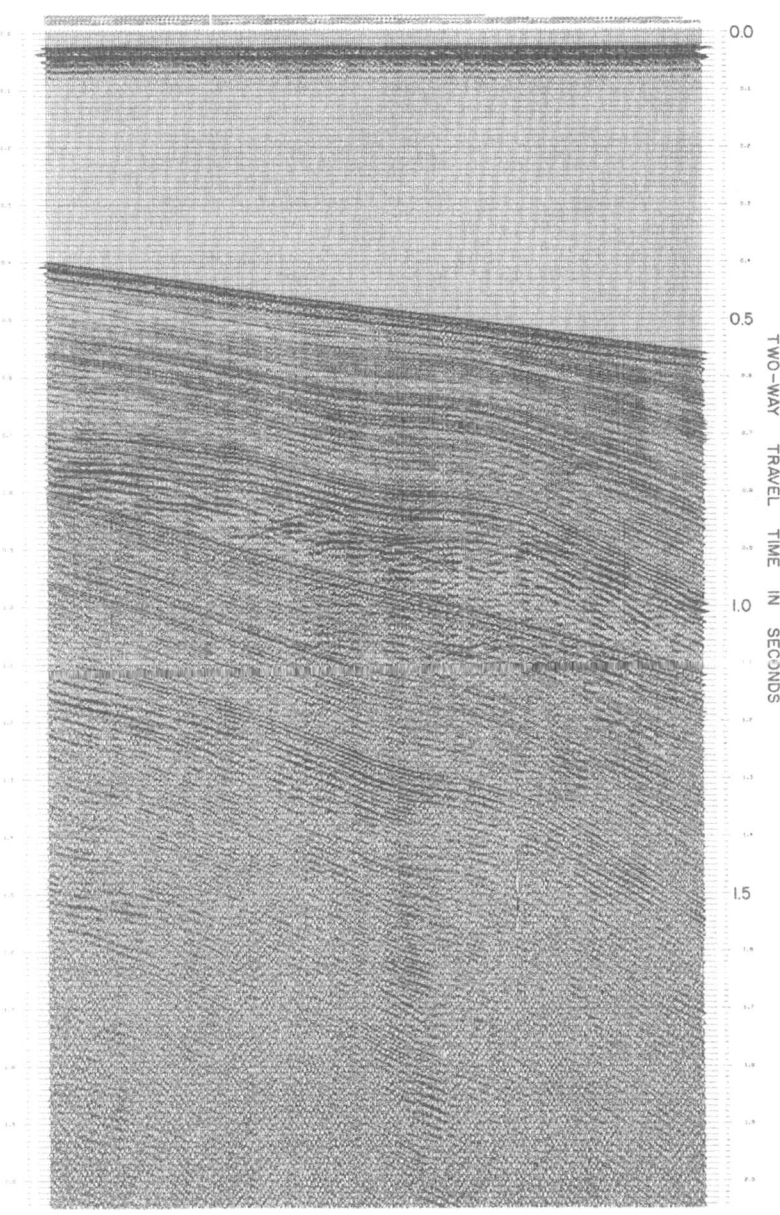

FIGURE 9.6 Near trace single channel display, equivalent to shipboard monitor record (2-sec record length, 1-msec sampling interval). Note presence of water-bottom and reflector multiples. Line 010 offshore Santa Barbara, California. (Courtesy of COMAP Geosurveys, Inc., Houston, Tex., processed by Sytech Corp., Houston, Tex.)

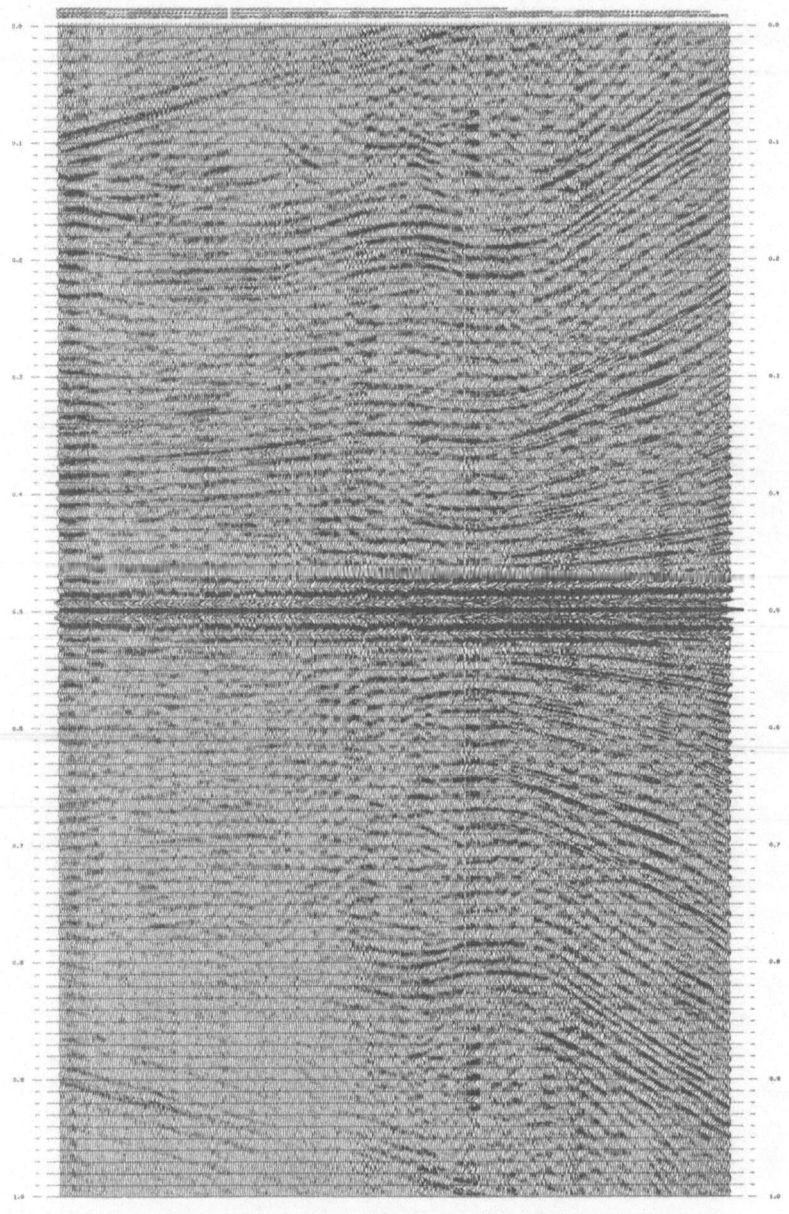

FIGURE 9.7 Display of autocorrelation pulse (± 500 msec) for near trace recordings, line 010, offshore Santa Barbara, California. (Courtesy of COMAP Geosurveys, Houston, Tex., processed by Sytech Corp., Houston, Tex.)

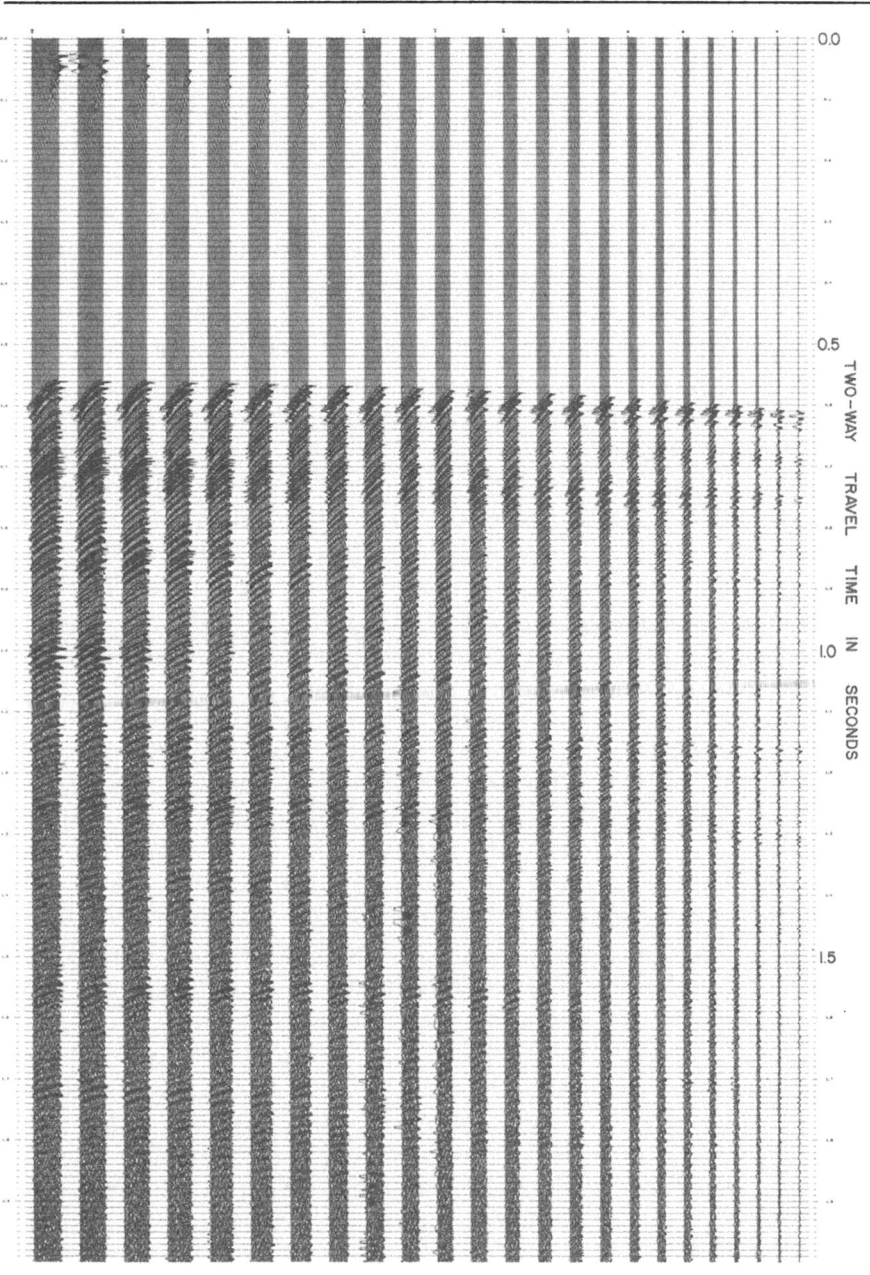

FIGURE 9.8 Display of multifold data (Line 010, offshore Santa Barbara, California) sorted into CDP order from CDP #1 (single-fold, *at right*) to CDP #24 (23-fold, *at left*). (Courtesy of COMAP Geosurveys, Inc., Houston, Tex., processed by Sytech Corp., Houston, Tex.)

which will eventually collapse the signal to a spike. The purpose of the operation is to reduce the length of a seismic pulse due to ringing.

4. The *sorting* of reflection traces into CDP gathers (fig. 9.8), is the assembly or grouping of seismic traces from common subsurface locations (midway between source and hydrophone). This step must be performed prior to the stacking of the multifold traces.

   Static corrections may be performed at this stage if the recording hydrophone cable was not maintained at constant depth, as in the use of the FLAIR technique (Patent Fairfield Industries, Inc.). While rarely performed on marine seismic reflection data, the application of static corrections is a routine step when working with land data, in order to remove the effects of topography, where it is referred to as elevation correction.

5. *Velocity analyses* are performed to obtain RMS (root mean square) velocities with depth by calculating arrival delay times of select reflecting horizons, caused by the increased distance along the hydrophone receiver cable away from the seismic source. This delay in arrival time is referred to as normal moveout (NMO). The number of velocity measurements is dependent upon suspected lateral variations in velocity for a particular survey area, and can be computed automatically by the computer program (figs. 9.9 and 9.10), or selected by visual inspection of constant velocity panels (figs. 9.11 and 9.12). The selected velocity function (by either of the above methods) is then applied to the CDP gathers.

6. *Velocity-corrected CDP gathers* are next generated (fig. 9.13). The selected velocity function (from figs. 9.9–9.12) was 4820 ft/sec (1469 m/sec) for the beginning of the records (time 0.0) and 7300 ft/sec (2225 m/sec) at 2.0 sec. This operation corrects for (removes) the effect of NMO time delays. The data (fig. 9.13) have also had early arrival water bottom signals muted (removed).

7. Figure 9.14 is a display of the *CDP stacked section*, corrected for NMO and velocities. A comparison with figure 9.6 shows the improvement of the multifold stacked data over the single trace section at this stage of the processing sequence.

8. *Gain recovery* of data with respect to time (depth) is next performed (fig. 9.15) to boost signal amplitudes, which have been lost through spherical spreading of the acoustic signal as it propagated away from the seismic source and reflectors, and to compensate for any losses caused by the attenuating effects of the signals as they travel through the water column and underlying sediments. The effect of gain recovery is illustrated by a comparison to figure 9.13.

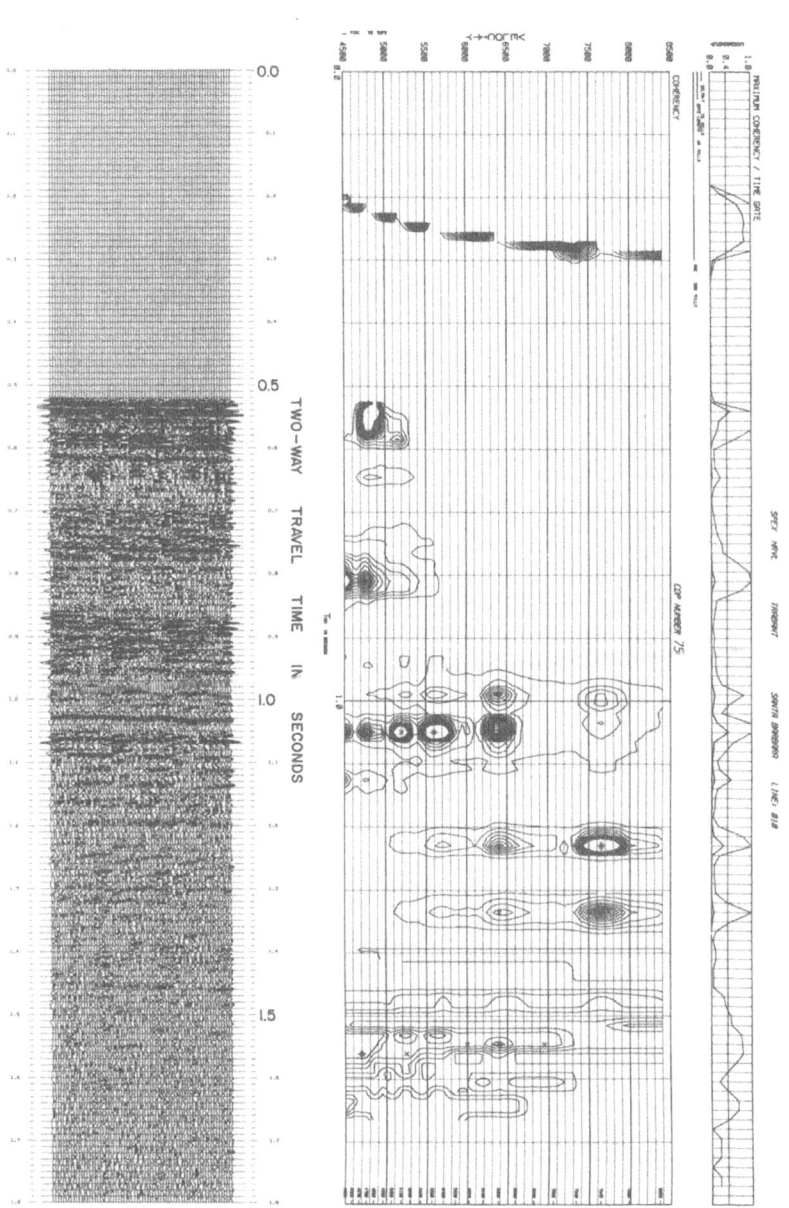

FIGURE 9.9 Display of computer-picked velocity analysis (contoured values *in right panel*), and resulting corrected profile (*left panel*) for CDP #75, Line 010, offshore Santa Barbara, California. (Courtesy of COMAP Geosurveys, Inc., Houston, Tex., processed by Sytech Corp., Houston, Tex.)

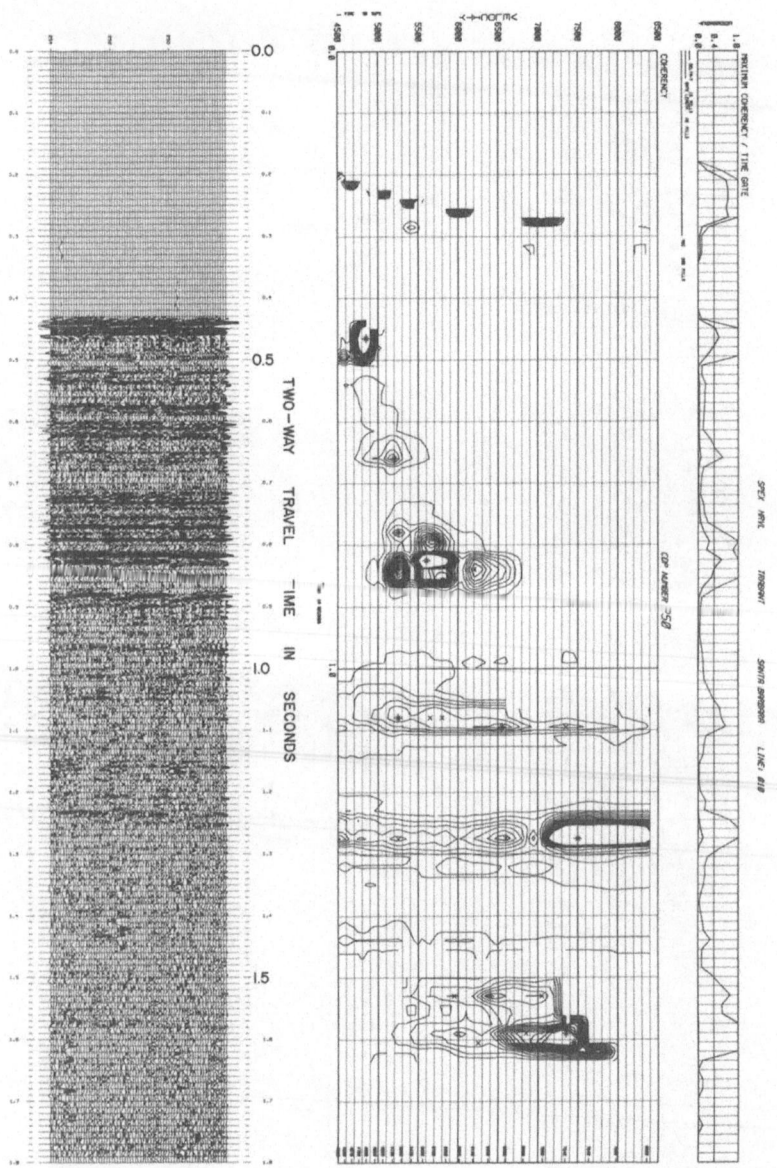

FIGURE 9.10 Display of computer-picked velocity analysis (contoured values *in right panel*), and resulting corrected profile (*left panel*) for CDP #250, Line 010, offshore Santa Barbara, California. (Courtesy of COMAP Geosurveys, Inc., Houston, Tex., processed by Sytech Corp., Houston, Tex.)

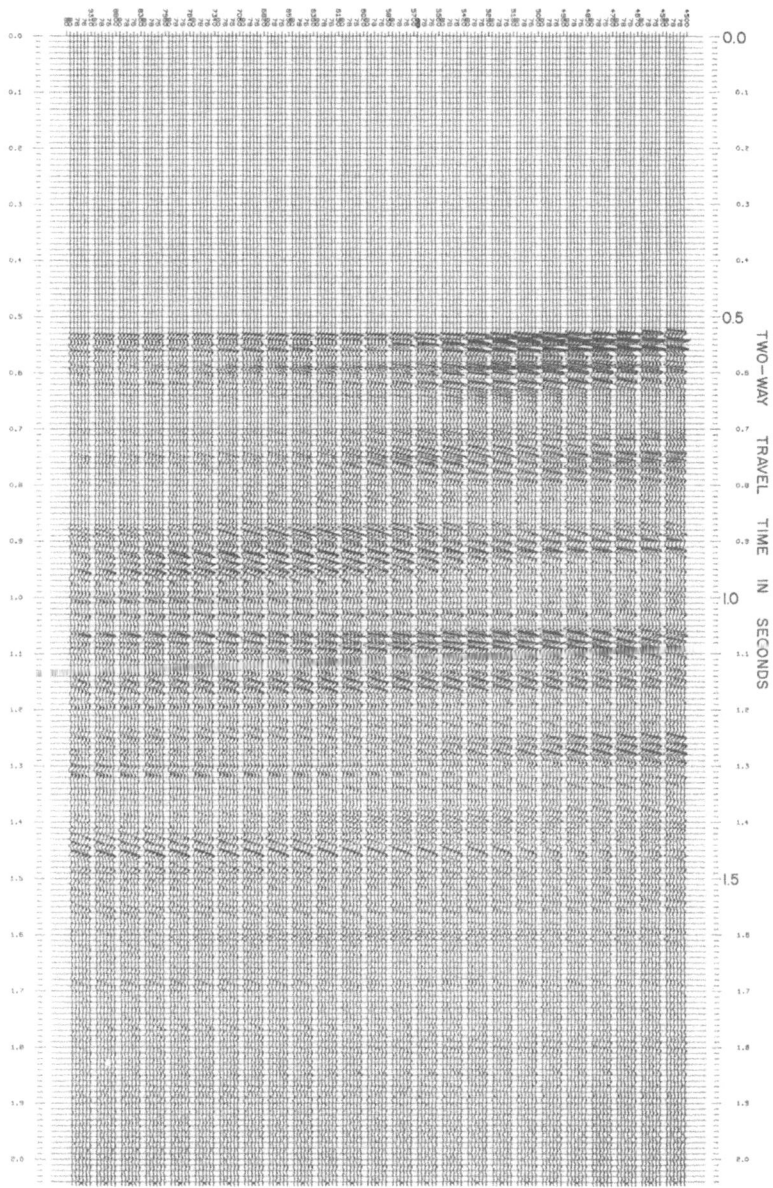

FIGURE 9.11 Constant velocity stacks for CDP #75; velocities range from 4500 ft/sec (1371 m/sec) *at right* to 9310 ft/sec (2838 m/sec) *at left*. Note coherence of reflectors and multiples at different velocities, Line 010, offshore Santa Barbara, California. (Courtesy of COMAP Geosurveys, Inc., Houston, Tex., processed by Sytech Corp., Houston, Tex.)

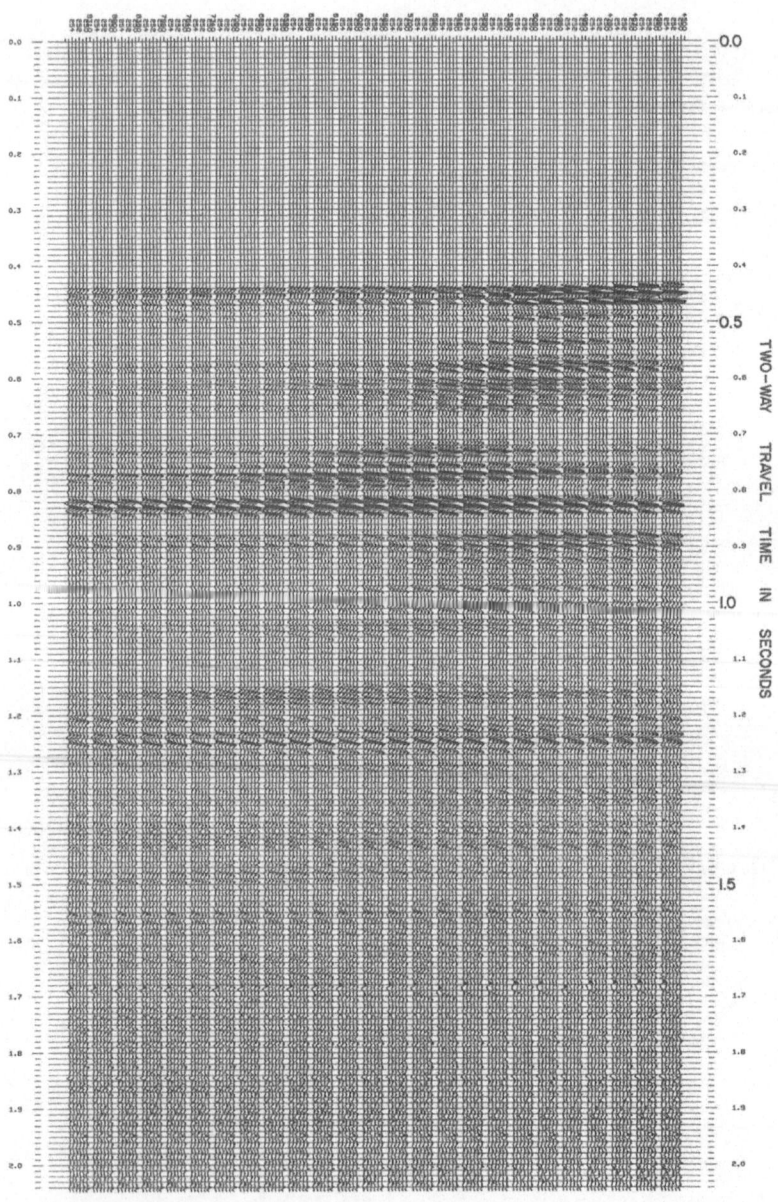

FIGURE 9.12 Constant velocity stacks for CDP #250; velocities range from 4500 ft/sec (1371 m/sec) *at right* to 9310 ft/sec (2838 m/sec) *at left*. Note coherence of reflectors and multiples for different velocities. Line 010, offshore Santa Barbara, California. (Courtesy of COMAP Geosurveys, Inc., Houston, Tex., processed by Sytech Corp., Houston, Tex.)

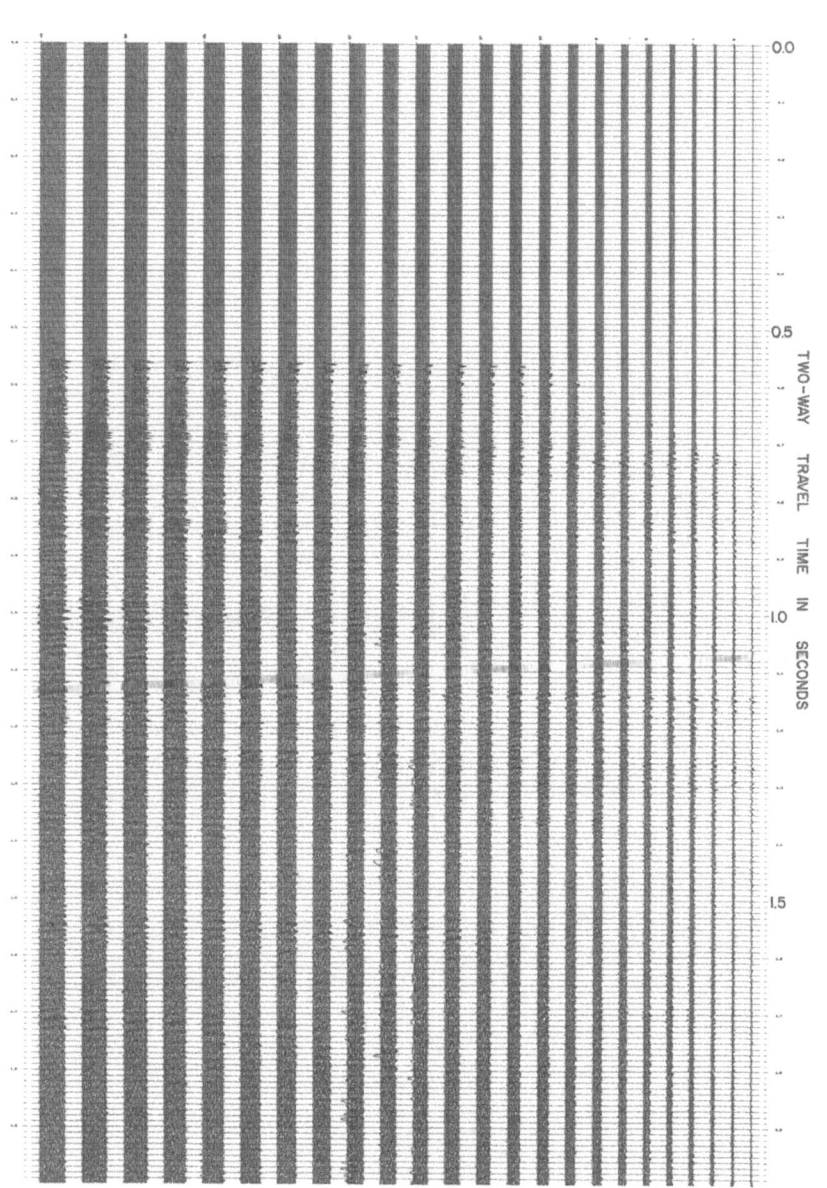

FIGURE 9.13 CDP gathers corrected for normal moveout velocities (beginning velocity: 4820 ft/sec [1469 m/sec], ending velocity: 7300 ft/sec [2225 m/sec]), and early arrival and water-bottom muting (cancellation) applied. Line 010, offshore Santa Barbara, California. (Courtesy of COMAP Geosurveys, Inc., Houston, Tex., processed by Sytech Corp., Houston, Tex.)

FIGURE 9.14 Multifold (24) CDP stack of Line 010, after NMO velocity corrections, offshore Santa Barbara, California. (Courtesy of COMAP Geosurveys, Inc., Houston, Tex., processed by Sytech Corp., Houston, Tex.)

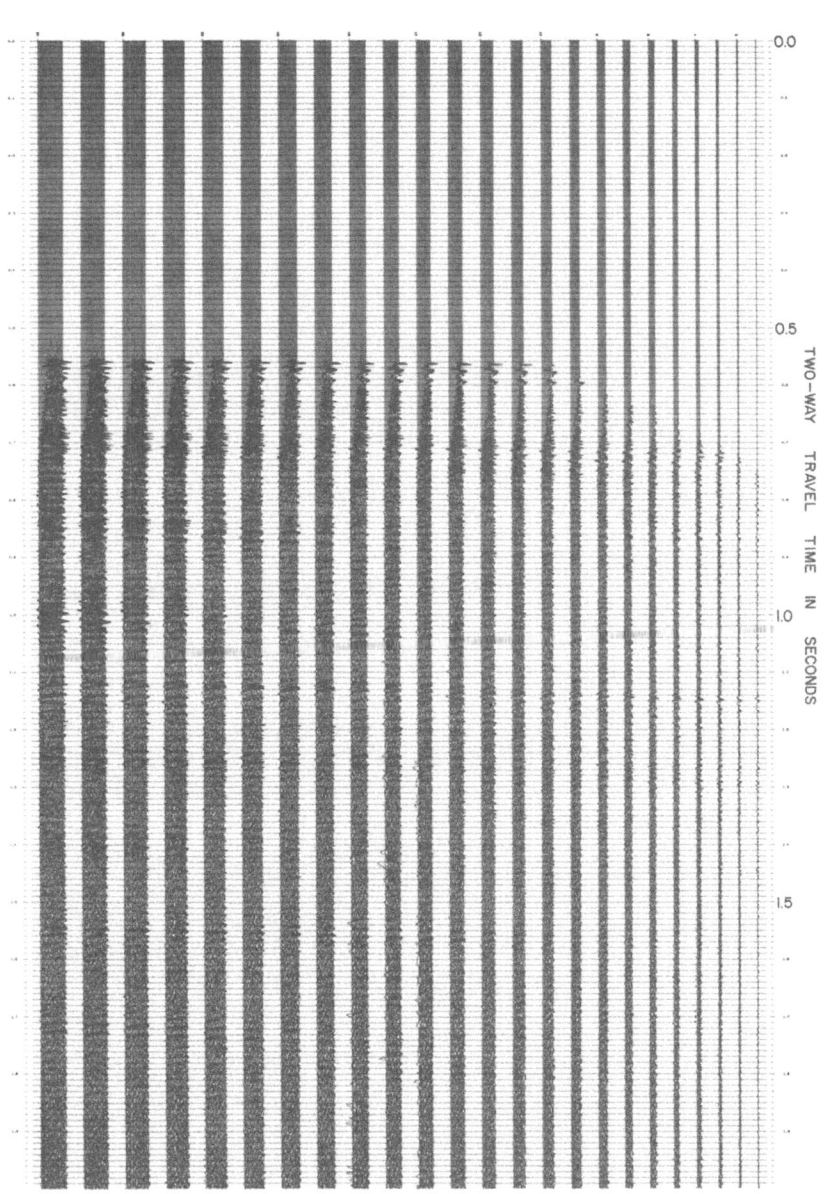

FIGURE 9.15 CDP gathers following stack (fig. 9.14), with additional gain recovery (T × V) processing to boost amplitude as a function of depth, Line 010, offshore Santa Barbara, California. (Courtesy of COMAP Geosurveys, Inc., Houston, Tex., processed by Sytech Corp., Houston, Tex.)

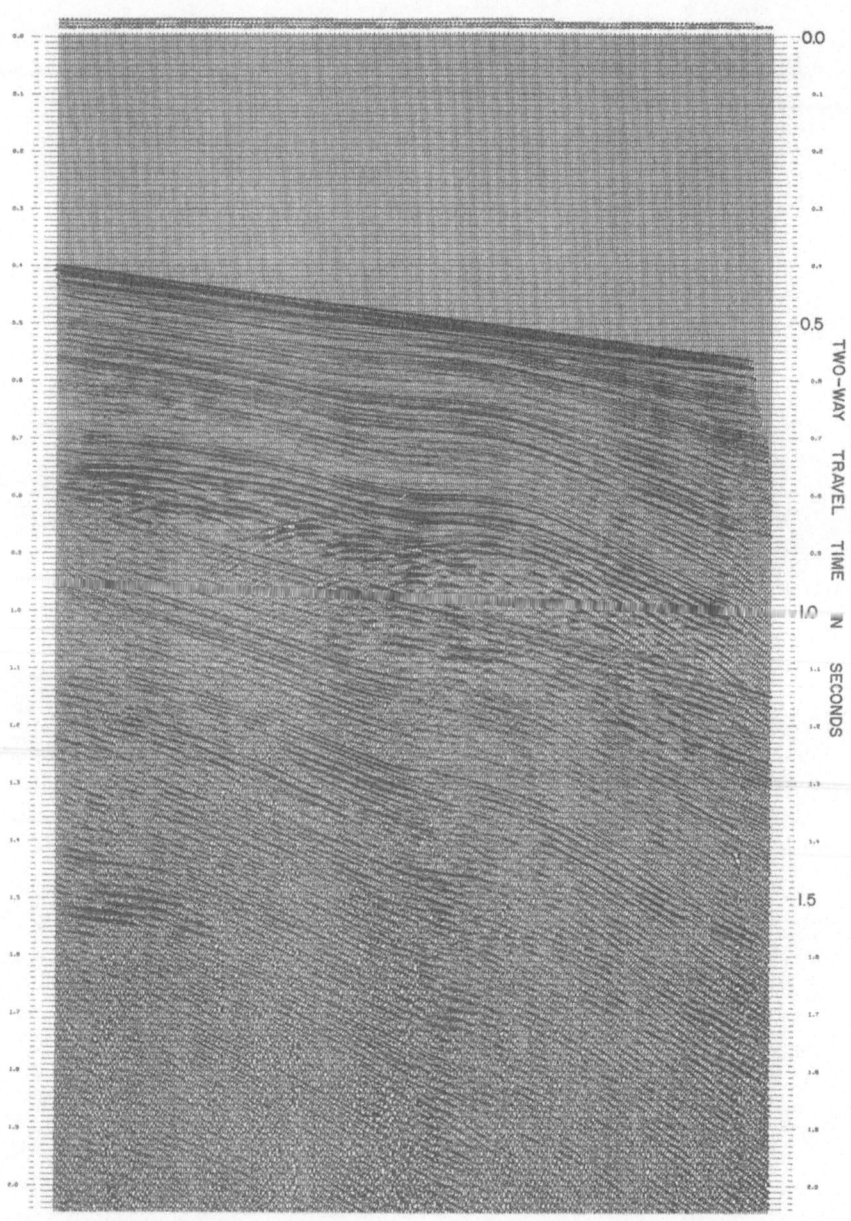

FIGURE 9.16 Multifold (24) CDP stack with gain recovery processing, Line 010, offshore Santa Barbara, California. (Courtesy of COMAP Geosurveys, Inc., Houston, Tex., processed by Sytech Corp., Houston, Tex.)

9. The data is again displayed (fig. 9.16) after both gain recovery and the application of a *predictive deconvolution* process (applied to a 30 msec window throughout the entire record length). This step spikes the seismic pulse to attenuate the effect of water-bottom multiples.

10. Further processing with a larger predictive deconvolution window (60 msec) is performed to collapse the seismic signal and remove the effects of the bubble pulse and ghost (fig. 9.17).

11. The *multifold stack* produced after the above processing is displayed in figure 9.18. The improved results are now clearly evident when compared to figures 9.6, 9.14, and 9.16, as the multiples are barely detectable.

12. A second autocorrelation of the data is performed (fig. 9.19) at this stage to assess improvements and help in the selection of further operating windows.

13. The selection of appropriate *filter* limits is simplest if done on a number of displays of processed data using different filter settings (fig. 9.20). Optimum filter settings with respect to depth may be selected on the basis of their visual display to enhance the signal-to-noise ratio. The applied filter settings (fig. 9.20 from right to left) are:

low cut	high cut	*(at 6 dB/slope)*
Hz	Hz	
none	none	
16	63	
22	85	
28	113	
38	151	
52	201	
68	269	

14. Following the selection of optimum filter limits, the data are again stacked (fig. 9.21) with filtering applied (bandpass: 52–201 Hz). This essentially completes the usual processing sequence for HRG multifold data. If, however, the survey profile contains strongly dipping reflectors, it is advantageous to take one more processing step.

15. Figure 9.22 shows the display of Line 010 upon application of *wave migration* processing. As stated in chapter 2, this process compensates for the effects of dip, by fitting the data (reflectors) into their true spatial locations. The effect of migration on these data is quite

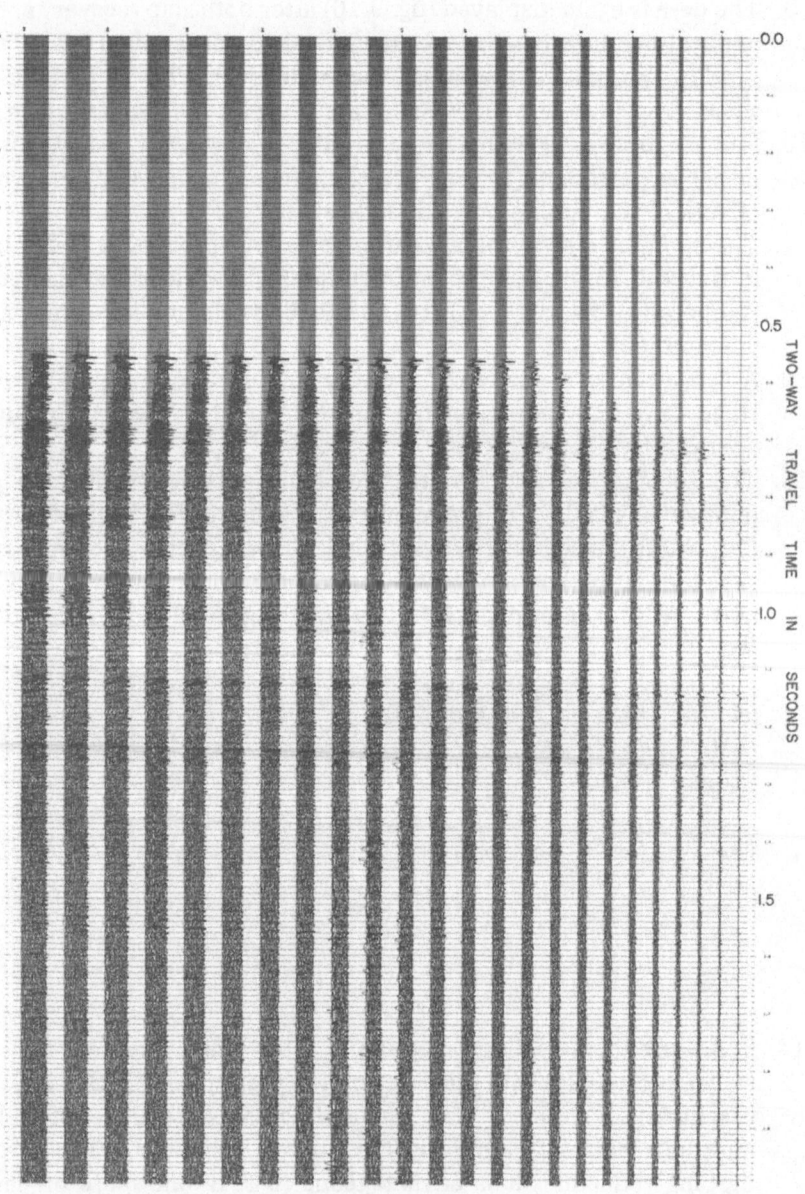

FIGURE 9.17 CDP gathers with predictive deconvolution processing applied to attenuate source bubble pulse. Operator window length of 60 msec. Line 010, offshore Santa Barbara, California. (Courtesy of COMAP Geosurveys, Inc., Houston, Tex., processed by Sytech, Inc., Houston, Tex.)

FIGURE 9.18 Multifold (24) CDP stack with predictive deconvolution processing applied (fig. 9.17), Line 010, offshore Santa Barbara, California. Note improvement (further reduction of multiples) over previous stacks. (Courtesy of COMAP Geosurveys, Inc., Houston, Tex., processed by Sytech Corp., Houston, Tex.)

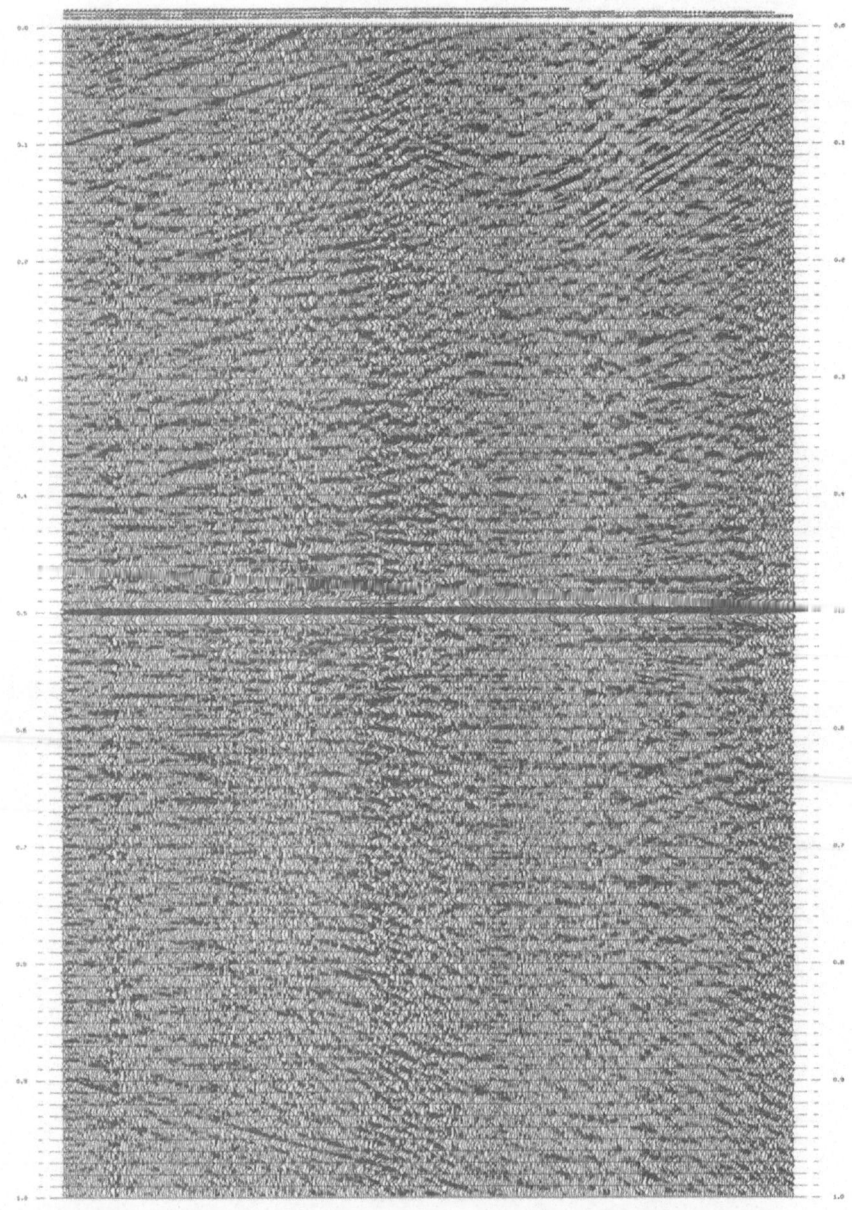

FIGURE 9.19 Display of autocorrelation pulse (± 500 msec) from stacked multifold section (fig. 9.18), Line 010, offshore Santa Barbara, California. (Courtesy of COMAP Geosurveys, Inc., Houston, Tex., processed by Sytech Corp., Houston, Tex.)

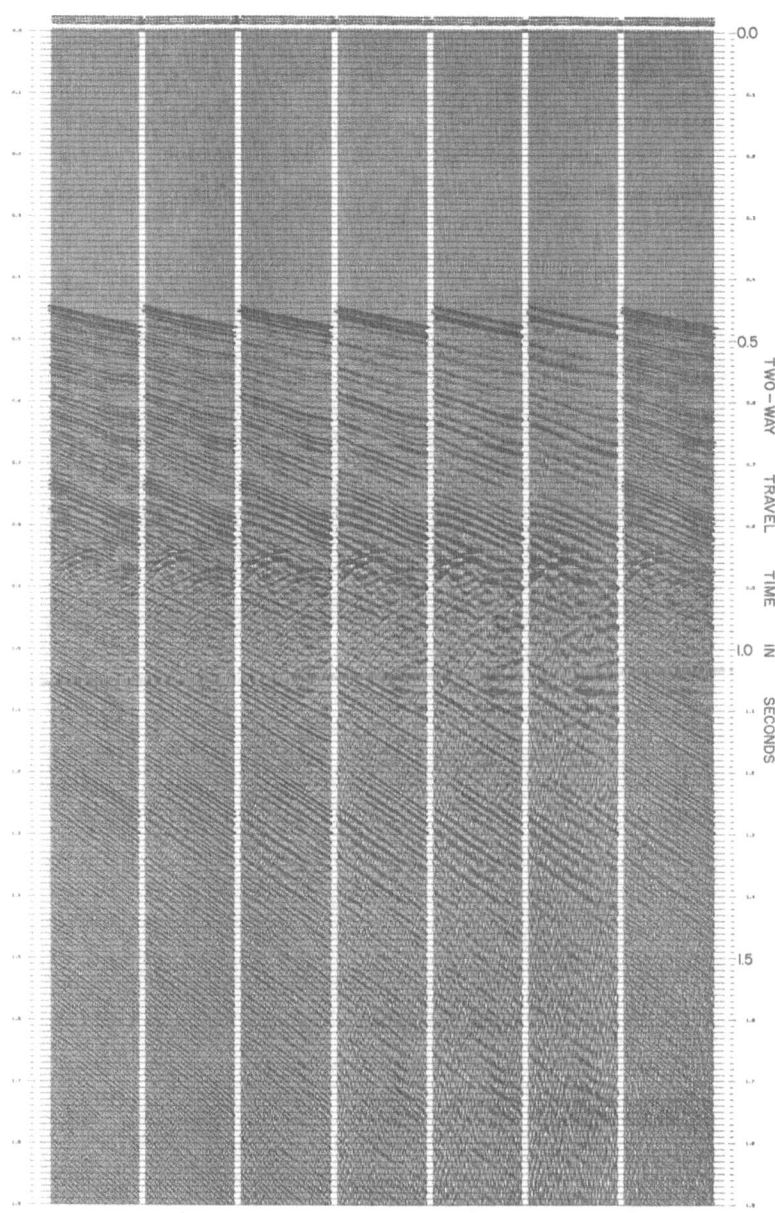

FIGURE 9.20 Display of individual filter panels for CDP's 176–235 (Line 010, offshore Santa Barbara, California), following the stack of figure 9.18. Panel at right has no filter applied; note variation of data appearance with applied filters. (Courtesy of COMAP Geosurveys, Inc., Houston, Tex., processed by Sytech Corp., Houston, Tex.)

FIGURE 9.21 Multifold (24) CDP stack of Line 010 with filter applied (band pass: 52–201 Hz, with 151 msec window operator), offshore Santa Barbara, California. (Courtesy of COMAP Geosurveys, Inc., Houston, Tex., processed by Sytech Corp., Houston, Tex.)

FIGURE 9.22 Multifold (24) CDP stack of Line 010 following application of wave equation migration processing (compare to fig. 9.21), offshore Santa Barbara, California. (Courtesy of COMAP Geosurveys, Inc., Houston, Tex., processed by Sytech Corp., Houston, Tex.)

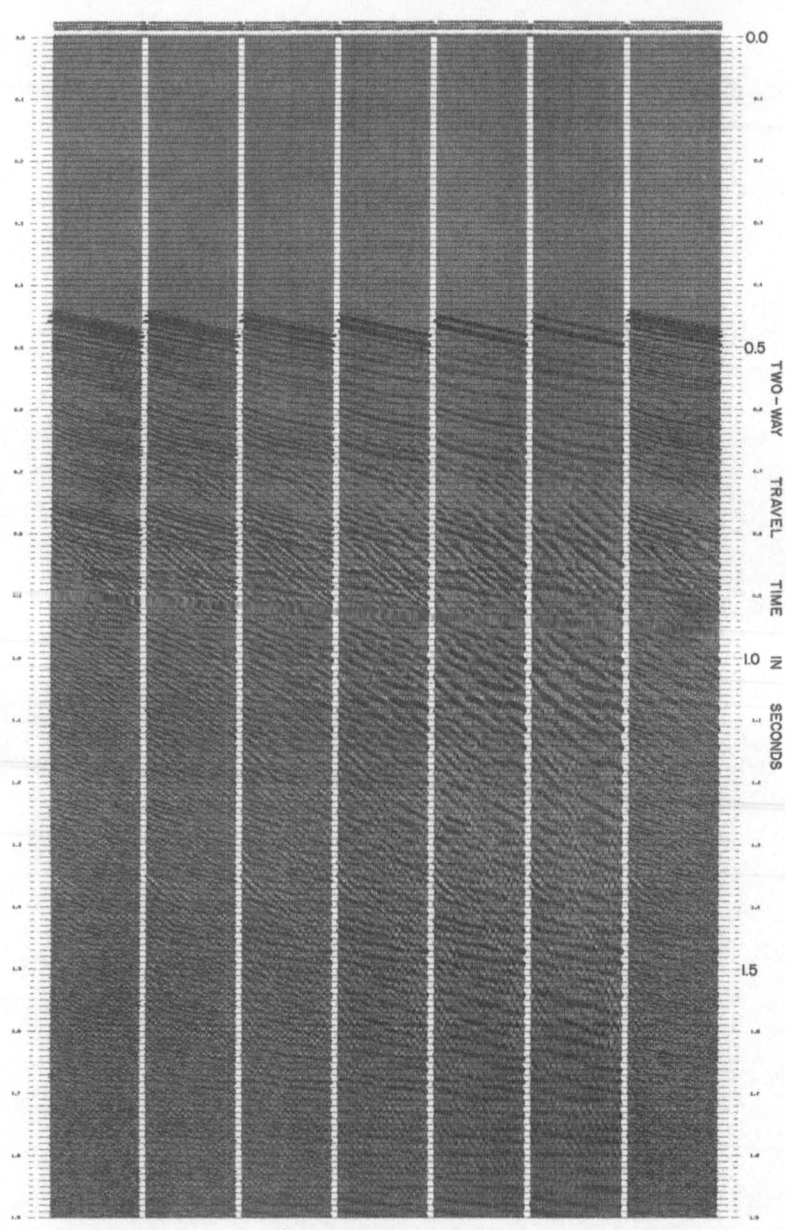

FIGURE 9.23 Display of individual filter panels for CDPs 176–235 (Line 010, offshore Santa Barbara, California) following migration (fig. 9.22); panel at right has no filtering. (Courtesy of COMAP Geosurveys, Inc., Houston, Tex., processed by Sytech Corp., Houston, Tex.)

pronounced (compare to figure 9.21), with the generation of inverse artificial diffraction patterns at depth.

16. An additional selection of filters (fig. 9.23) prior to production of the final display again helps increase the signal-to-noise ratio. The filter settings are the same as for step 13.

17. The *final product* with additional filtering applied to the wave migrated section is shown in figure 9.24. This should be compared to the initial input (fig. 9.6) where multiple interference, signal amplitude decay with depth, and incorrect dips would not have produced identical interpretations. This display is commonly referred to as a *structural section*.

18. Figure 9.25 is a display of the previous section (fig. 9.24) after the application of relative *amplitude processing*. The display (referred to as an *RAP section*) permits the identification of *bright spots* and other amplitude anomalies, an important step in the interpretation of geoengineering hazards before drilling operations.

OPERATIONAL CONSIDERATIONS

A large number of items must be considered in the proper deployment and operation of a multifold acquisition system. All sensor deployment parameters should be checked with the data processor, prior to survey operations, to ensure straightforward postsurvey data processing. The recording format for digital seismic reflection data on magnetic tape should be in line with standards, such as those set by the Society of Exploration Geophysicists (4).

The correct source/sensor geometry should be established prior to survey operations, as well as monitored, and recorded during the course of the survey. This includes the hydrophone depth, and their angle to the vessel, which is monitored by radar or special cable sensors. The speed of the survey vessel is critical since combined with the source firing rate, it will establish the overall system geometry, a highly critical element in later data processing. Basically, the correct firing rate is established as a function of the survey vessel's speed, recorded record length, and group interval, such that the source is fired to provide a common depth point every half group interval distance.

The correct deployment of the hydrophone cable involves time and patience, in order that the system be made neutrally buoyant, and all depth indicators be properly calibrated to indicate the correct streamer depth.

FIGURE 9.24 Final migrated structural section of Line 010 (offshore Santa Barbara, California), with final filter (52–201 Hz) applied. (Courtesy of COMAP Geosurveys, Inc., Houston, Tex., processed by Sytech Corp., Houston, Tex.)

FIGURE 9.25 Relative amplitude display of Line 010, offshore Santa Barbara, California, following final processing. Note bright spots at 800 and 1260 msec. (Courtesy of COMAP Geosurveys, Inc., Houston, Tex., processed by Sytech Corp., Houston, Tex.)

Continuous monitoring of the hydrophone signals for noise caused by the presence of other vessels, fish, other seismic sources, rough seas, and 60 and 50 Hz signals, should all be done with an oscilloscope and monitor camera, at the same time frequent checks of the recording system are being performed.

Criteria as to the maximum noise levels and numbers of inoperative groups should be established and adhered to prior to a survey.

Operations in shallow waters or within areas of heavy traffic should be prepared for remedial procedures, such as the rapid raising of the streamer to avoid its grounding and destruction by raising or lowering the cable with depth controllers, or by changes in the survey vessel's speed.

INTERPRETATION

While basically an art, based upon experience and imagination, the interpretation of multifold seismic reflection records involves a detailed knowledge of the effects of the processing sequence performed on the data. Comparisons of nonprocessed analog sections (e.g., near trace monitor records, as in fig. 9.6) with processed ones (fig. 9.24 and 9.25) provides excellent experience in this respect.

Essential to an interpretation are the specific objectives of the survey, whether for foundation stability and penetration of shallow subbottom depths, or the detection of potential hazards to drilling operations. The depth of the objectives being different may require different enhancements of the data for the particular objectives. Thus, when drilling operations are involved it is essential to have HRG data of maximum penetration, as well as relative amplitude displays (RAPs), to detect faults and near-surface gas accumulations (bright spots). An excellent set of illustrative HRG seismic sections may be found in the AAPG *Picture and Work Atlas* (5).

CONCLUSIONS AND FUTURE TRENDS

Over the past decade the acquisition and processing of marine HRG seismic reflection data has followed a steady course, from the simple single channel analog system to the present-day equivalent of modern exploration seismic systems. Further developments and additional uses of the technique are envisioned for the future. While the major use for multifold HRG seismic reflection data has been in the determination of geoengineering hazards to the emplacement of structures on the sea floor, and drilling operations (depth of drill casing), future uses are expected in the assessment of shallow stratigraphic hydrocarbon traps, and toward the detailed geology of ocean basins and continental shelves.

General developmental trends include the use of more phone groups, varying cable depths, angles and group intervals, digitizing of data at the phones, the use of lighter weight fiber optic transmission cables, hard disc recordings and the use of shipboard processors for real-time data processing.

Ten
MARINE MAGNETOMETERS

INTRODUCTION

The inclusion of marine magnetometer systems as an integral part of high-resolution seismic surveys has become prevalent since the 1960s. The detection of ferromagnetic objects, on or near the sea floor along a survey vessel's track, is possible through the measurement of anomalies to the earth's total magnetic field caused by the object.

Typical features detected include: sunken ships, pipelines, communication cables, anchors and chains, unexploded ordnance (fig. 10.1), metallic debris and many other objects that could hinder the installation of structures on the sea floor. Magnetic anomalies are also associated with geological features such as faults, volcanic intrusions, and other structures that produce lateral variations of the magnetic susceptibility of underlying sediments and rocks. This technique has been used extensively in the exploration for minerals and hydrocarbons.

HISTORICAL BACKGROUND

The first documented use of the earth's magnetic field is attributed to the ancient Chinese with the discovery of the compass. This property was rediscovered during the twelfth century in Europe, and by the sixteenth century Joao de Castro had discovered the discrepancy between geographic and magnetic pole locations, as well as the secular variation of lines of magnetic flux over periods of hundreds of years. The fluxgate magnetometer was used during World War II for the detection of submarines, while the proton precession magnetometer was developed in the 1950s by Packard and Varian (1). The optical pumping type magnetometers (Rubidium and Cesium Vapor) were developed during the 1960s (2, 3) and provided greater sensitivity than either of the earlier types.

In view of the predominant use of the proton type magnetometer and its advantages over other types in the conduct of geohazard surveys, this chapter will only consider such systems as well as their use in tandem as gradiometers. An excellent review of the different systems may be found in geophysical textbooks (4, 5), and a publication on portable magnetometers by Breiner (6), as well as in the manuals supplied by magnetometer manufacturers.

THE EARTH'S MAGNETIC FIELD

The magnetic properties of the earth are still poorly understood, as the earth's magnetic field does not act as a single bar magnet dipole but rather

FIGURE 10.1 Unexploded ordnance located by marine magnetometer survey, Tobruk, Libya. (Courtesy of Jack Hudson, Cultural Resource Services, Inc., Seabrook, Tex.)

as a complex number of dipoles. If the earth had a single pole aligned from North to South, as a bar magnet, lines of magnetic flux would be emitted normal to the earth's surface at the poles and be horizontal at the magnetic equator. This is not the case, however, and results in the complex and changing magnetic field depicted in figure 10.2.

The field or flux lines of the earth's magnetic field are measured in units of Gauss (G), whereby convention in the study of geophysics:

$$1 \gamma = 10^{-5} \text{ G} = 10^{-5} \text{ Oe} = 10^{-9} \text{ Wb/m}^2 = 10^{-9} \text{ T}.$$

Magnetic anomalies, described in the field of high-resolution geohazard surveys, are given in units of gammas (γ) because of the greater convenience of the smaller unit. The total field strength map of the earth (fig. 10.2) illustrates the complexity of the magnetic flux distribution,

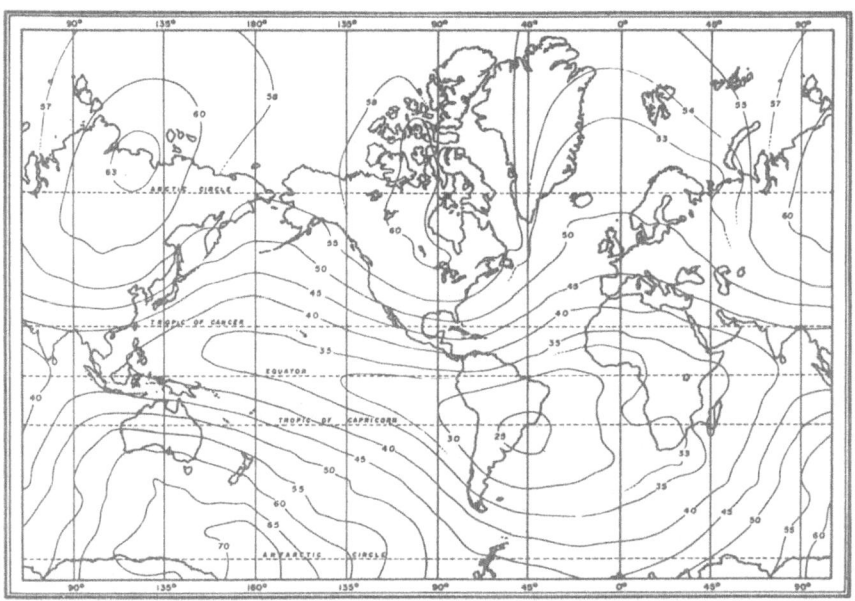

FIGURE 10.2 World map, depicting earth's total magnetic field. (Courtesy of EG&G Geometrics, Sunnyvale, Calif.)

which varies between approximately 25,000 γ at the magnetic equator, and 65,000 γ at the magnetic poles.

The value of the lines of magnetic flux, however, also vary with time, both of short and long duration, which are referred to as noise or interference when attempting to measure small local variations or anomalies. The variations are due to:

Large-scale regional effects, sometimes associated with regional mineral masses or mountains.

Secular variations, which are long-term (one hundred years and more) changes in the earth's magnetic field. For instance, these are responsible for the annual changes in magnetic variation noted on nautical charts and the wandering or changes in location of the magnetic poles over geological time spans. It may also be noted that magnetic reversals of the poles take place on this time scale.

Short-term variations on the order of seconds (micropulsations) to hours (diurnal changes) are the result of solar winds, which consist in a more or less constant flux of particles and electric currents emanating from the sun. These are associated with solar flares and are usually unpredictable.

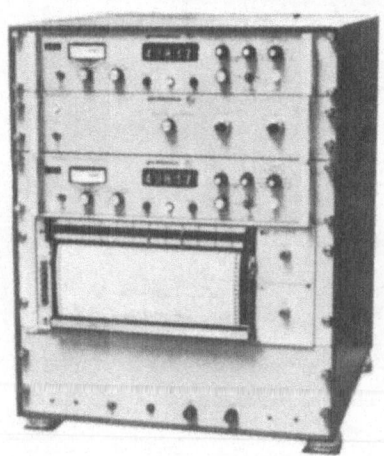

A

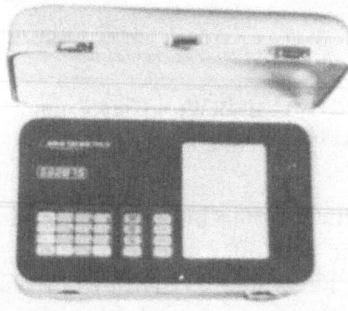

C

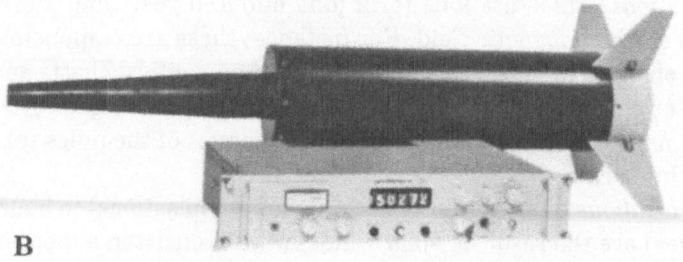

B

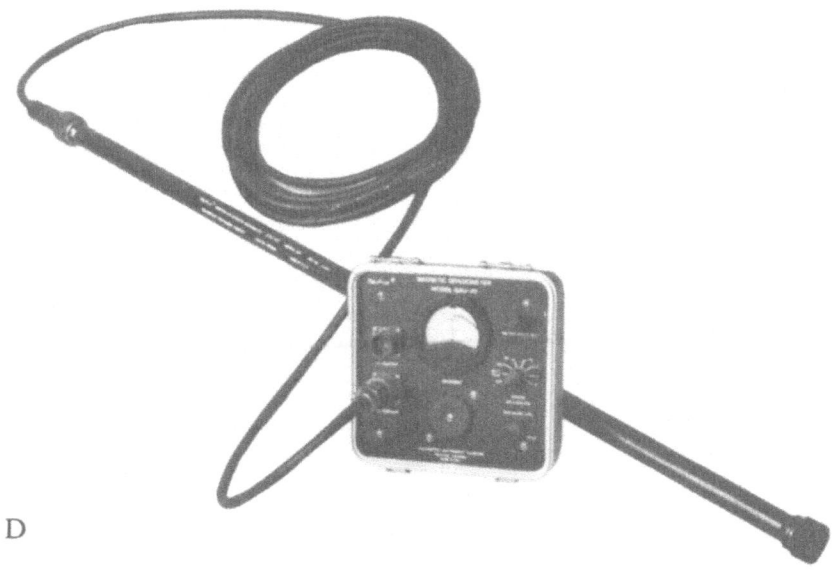

D

FIGURE 10.3 Photographs of a number of typical magnetometer sensor/recording systems: (A) Marine magnetometers mounted in tandem as gradiometer, (B) Magnetometer and tow-sensor "fish," (C) Portable digital magnetometer and recorder, (D) Portable magnetic gradiometer and sensor "stick." (A, B and C courtesy of EG&G Geometrics, Sunnyvale, Calif. D courtesy of Schonstedt Instrument Company, Reston, Va.)

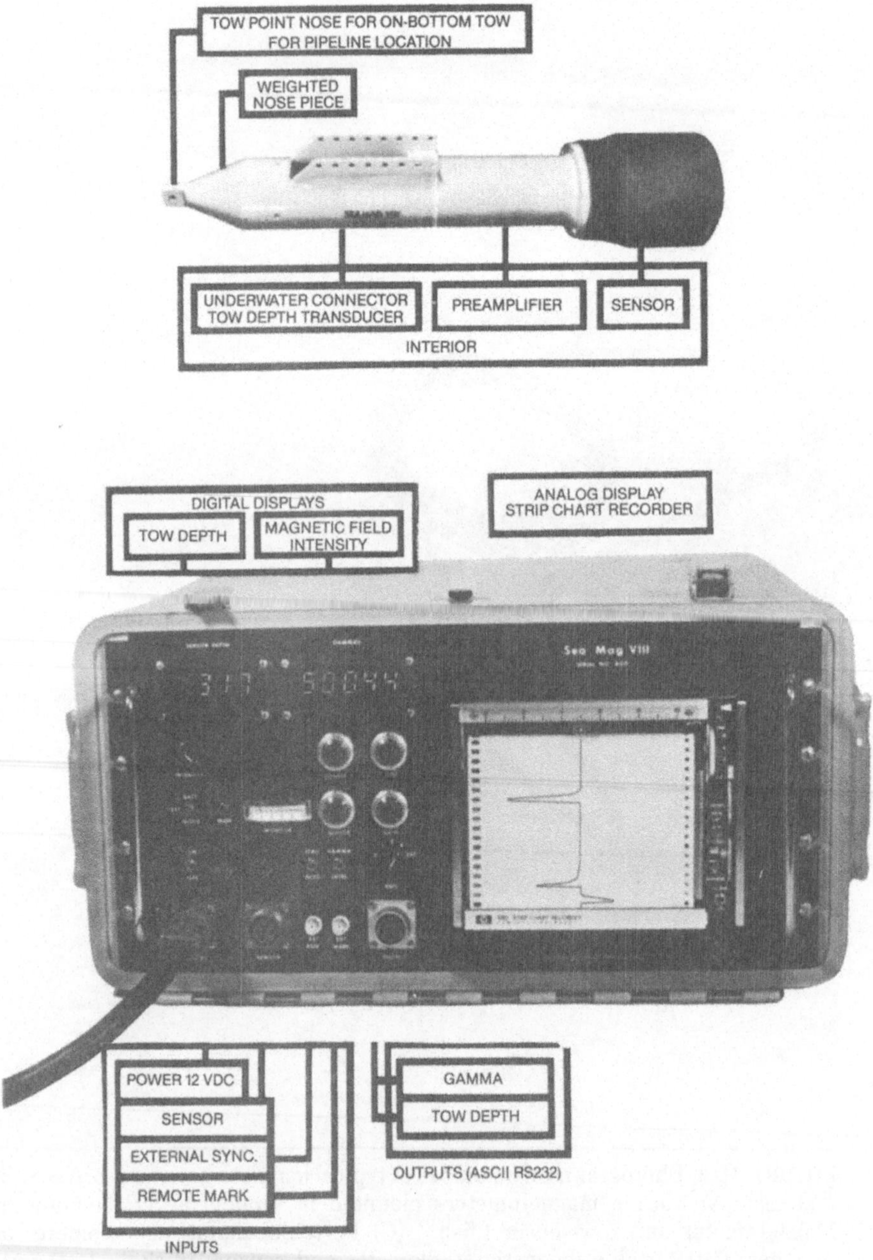

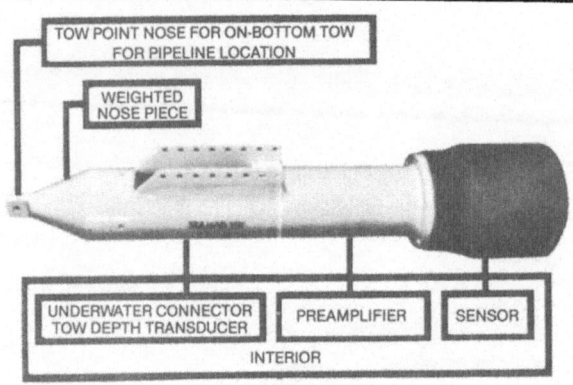

TOW POINT NOSE FOR ON-BOTTOM TOW
FOR PIPELINE LOCATION

WEIGHTED
NOSE PIECE

UNDERWATER CONNECTOR
TOW DEPTH TRANSDUCER

PREAMPLIFIER

SENSOR

INTERIOR

DIGITAL DISPLAYS

TOW DEPTH

MAGNETIC FIELD
INTENSITY

ANALOG DISPLAY
STRIP CHART RECORDER

Sea Mag VIII

POWER 12 VDC

SENSOR

EXTERNAL SYNC.

REMOTE MARK

INPUTS

GAMMA

TOW DEPTH

OUTPUTS (ASCII RS232)

FIGURE 10.4 Magnetometer sensor head (*top*), and recording unit (*bottom*) with block diagram of components, inputs and outputs. (Courtesy of Odom Offshore Surveys, Inc., Baton Rouge, La.)

The short-term variations may affect magnetometer readings by up to tens of gammas, and may only be accounted for and corrected by the use of gradiometer systems. The use of base stations near the survey site is not usually adequate for removal of these effects.

In addition, the effect of the electrojet system of the ionosphere along the magnetic equator may produce daily variations of up to 200 γ per day.

Shipboard-generated noises may also be detrimental to the use of magnetometers, and should be noted on the recordings when noticed; for instance, when generators are switched, or heavy loads turned on, noise spikes are sometimes generated and may be mistaken as anomalies. Radio transmissions may also produce noise on magnetic recordings.

PRINCIPLE OF OPERATION

The proton precession magnetometer provides a total magnetic field measurement at continuous sampling rates of up to half a second with a sensitivity of one γ. A typical unit consists of a shipboard electronics package, a tow cable, and a sensor head (figs. 10.3 and 10.4). The sensor consists of a cylindrical container for a hydrogen-bearing fluid, such as deionized water or kerosene, within which a wire coil is set for generating a magnetic field by passing an electric current. When a current is applied to the coil, a magnetic field is produced, causing the protons of the hydrogen fluid molecules to align with the generated magnetic field. Upon removal of this current the generated magnetic field collapses and the protons *precess* or realign themselves with the earth's magnetic field. In this process they generate a very weak alternating current in the coil for which the frequency is proportional to the earth's total magnetic field. This signal is independent of the orientation of the coil and thus very simple to operate by simply towing the sensor behind the survey vessel (fig. 10.5).

The proportionality constant that relates the frequency of precession to the earth's magnetic field is a well-known atomic constant, referred to as the gyromagnetic ratio of the proton (5). The signal level generated by the sensor is on the order of microvolts, and the frequency must be measured to an accuracy of approximately 0.04 Hz.

OPERATION

The proton precession magnetometer is referred to as an earth's field-determined component magnetometer. Such magnetometers are rela-

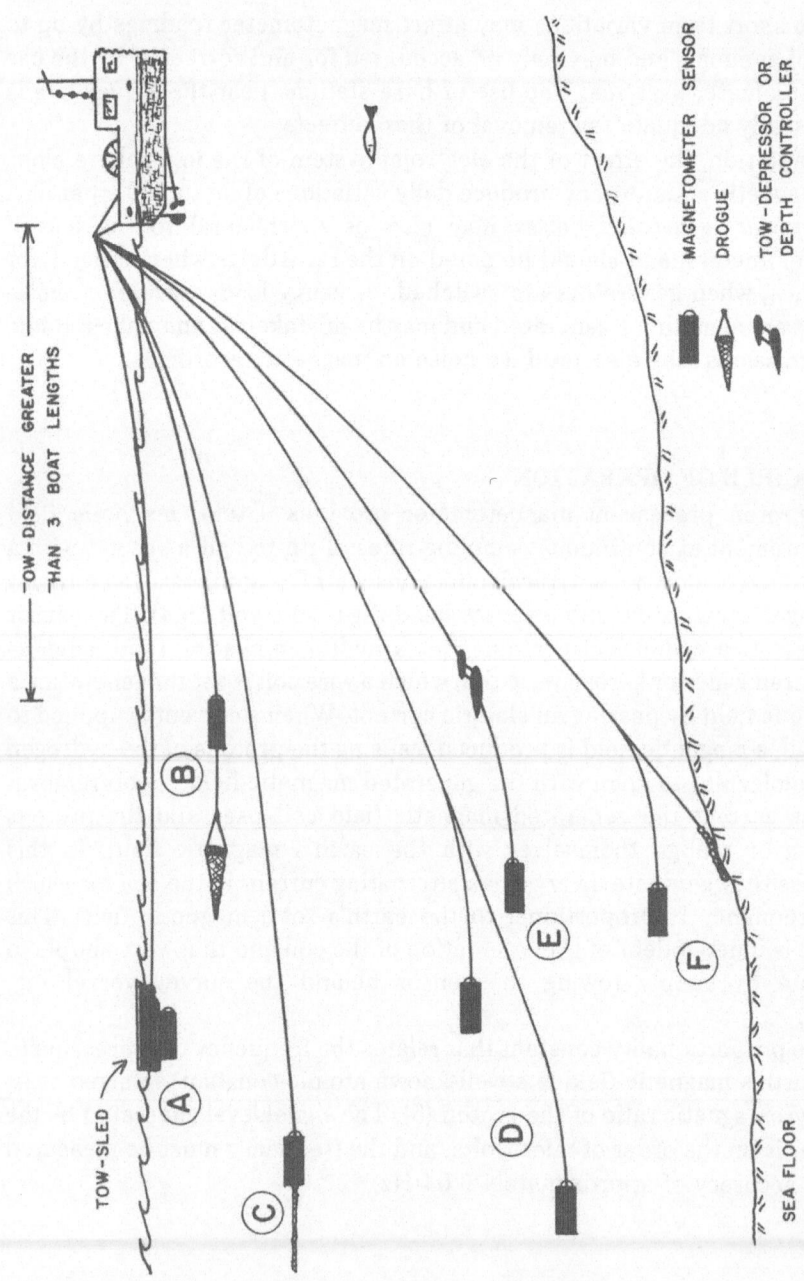

FIGURE 10.5 Diagram of survey vessel and a number of magnetometer sensor deployment methods: (A) surface tow on float, (B) near-surface tow with drogue, (C) subsurface tow with rope drogue, (D) subsurface tow of two sensors as gradiometer, (E) deep tow using tow-depressor, (F) deep tow using bottom-drag of chain to maintain constant height off sea floor. (By the author.)

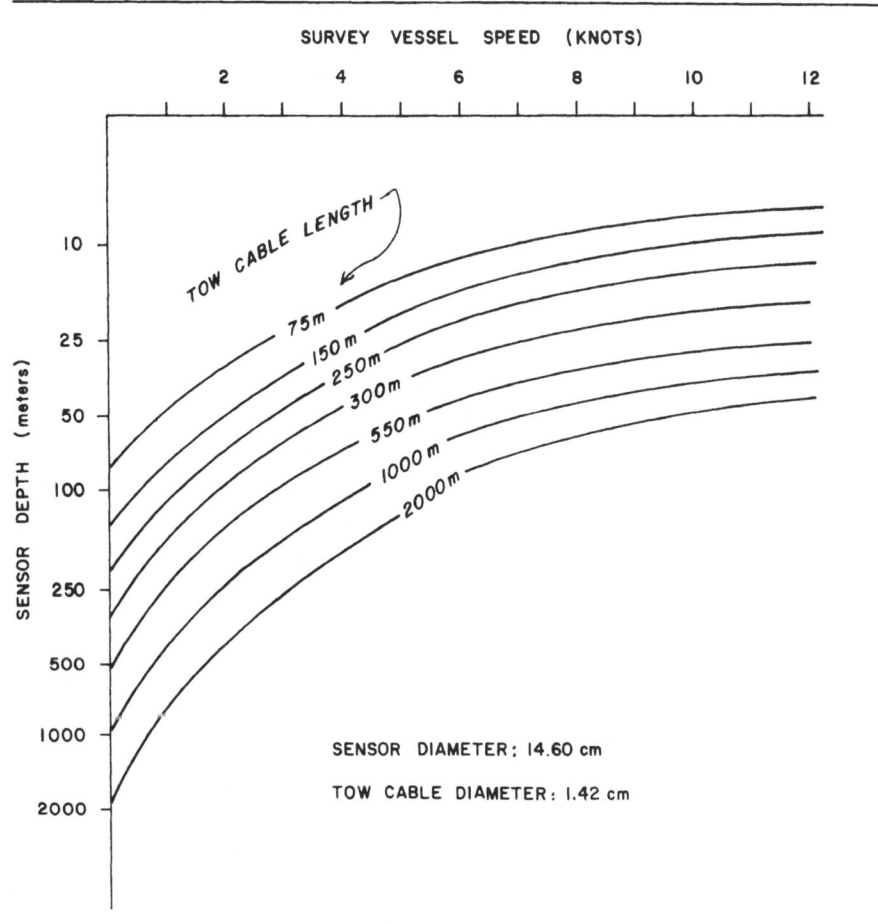

FIGURE 10.6 Graph of magnetometer sensor tow-distance, versus survey vessel speed and sensor cable length (for specific cable and sensor head). (Modified from Breiner, S., 1975, Marine magnetics search: EG&G Technical Report #7, EG&G Geometrics, Sunnyvale, Calif.)

tively simple in operation, containing no moving parts, and provide relatively high-resolution measurements in the field.

The signal, however, becomes degraded with the presence of large magnetic field gradients greater than about 600 γ/m, and may not be operated near alternating currents and electrical power sources (e.g., ship's generator). Thus, the sensor head must be towed behind the survey vessel at a distance on the order of several boat lengths (see fig. 10.6 for appropriate distance). The cable must, of course, be of sufficient strength to endure the rigors of the expected tow operations at sea while providing the necessary conductivity for both signals, and is usually a triple or quadruple conductor, with underwater conductor plug attachments at each end. Slip rings on the winch, which allow continuous electrical operations while the tow cable is spooled in and out, are generally not necessary as the cable is payed out an appropriate length at the commencement of a survey operation and remains fixed until completion.

If the magnetometer survey is to be conducted in shallow water, a buoy may be attached to the sensor head, or a floating tow cable used, to prevent it from dragging on the bottom as well as to maintain a near constant height above the sea floor (fig. 10.5).

The necessary electronics may consist of separate components (signal generator, tone burst generator, oscilloscope, amplifier and frequency counter), or a prepackaged unit such as shown in figures 10.3 and 10.4. Advances in electronics over the past decade have produced high-density integrated circuits allowing a reduction in size of the typical seagoing magnetometer electronics of many orders of magnitude (fig. 10.3c).

GRADIOMETERS

Gradiometers are simply magnetometers, usually a pair, attached in tandem with a known spacing (vertical or horizontal) (fig. 10.3a). Thus the difference in readings between the two sensors, divided by the distance separating the units, provides a gradient or rate of change of the magnetic lines of flux.

For marine search work a separation, on the order of 150 m between two towed proton precession units, provides a measurement of the local magnetic variation by removing unwanted effects caused by long- and short-term magnetic variations (as with magnetic storms and diurnal variations). Such data also allow (with certain assumptions) the quantitative determination of the depth and/or size of an object (6).

INTERPRETATION

Magnetic field disturbances caused by ferromagnetic objects on the sea floor or small geological features produce anomalies that must be distinguished from those caused by noise (fig. 10.7). The *amplitude* of an anomaly is proportional to the the distance between the sensor and the object (6). By manually smoothing the recorded curves (fig. 10.7) to remove the effects of short- and long-term noises, one is left with

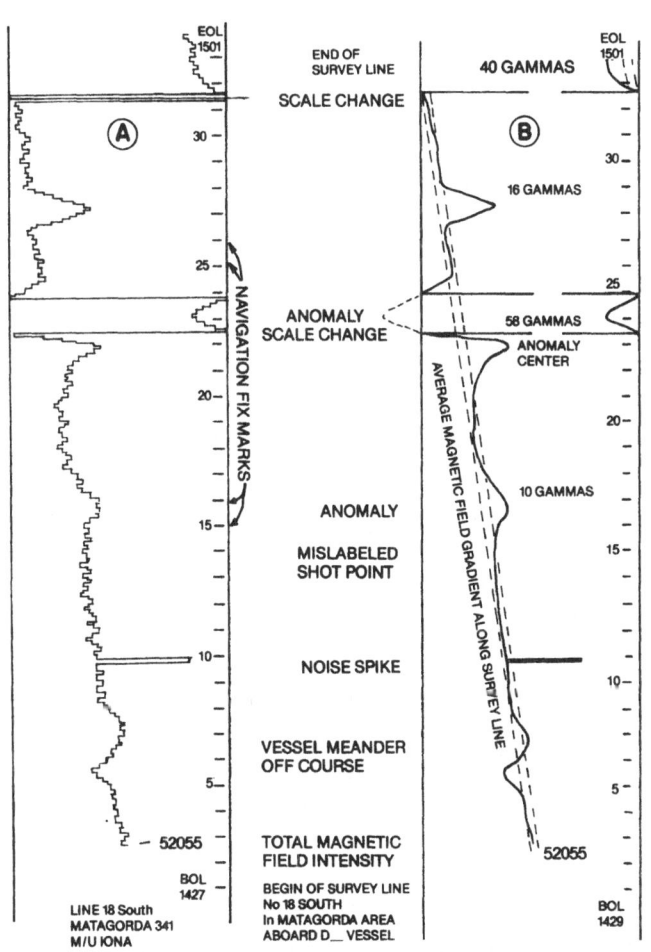

FIGURE 10.7 Sample magnetometer record (A), with interpretation of features (B). (Drawing by the author.)

unidentified irregularities referred to as *anomalies*, which may be due to the presence of a ferromagnetic object. Care and experience are required in this process, as well as knowledge of the size of detectable objects and their signatures.

The removal of the earth's *regional gradient* (fig. 10.7) is not generally necessary for the detection of anomalies. Those anomalies of concern in the conduct of geohazard surveys are generally characterized by a symmetrical pattern (fig. 10.7), while the amplitude varies according to: direction (orientation) of the earth's field, configuration of the source of the anomaly, and most important the distance between the source of the anomaly and the sensor.

The shape and amplitude of an anomaly permit estimations as to the size of the ferromagnetic object, and location, which may not necessarily be below the vessel's track but adjacent to it. Precise location of an object producing an anomaly may require a detailed survey grid spacing and additional survey lines. The relative amplitudes of anomalies, based upon known features such as a 12 in. pipeline within a survey area, permit an estimation of the size of other features.

The correlation of magnetic anomalies with other high-resolution sensor data, such as side scan sonar, frequently permit the positive identification of features such as a pipeline or sunken vessel. This correlative process also applies in reverse, whereby features observed on other sensors should be cross-checked against the magnetometer data. Thus, if an object or target observed on side scan does not produce a magnetic anomaly, one may assume that the object has no ferromagnetic material. If, on the other hand, a magnetic anomaly is found to have no correlative feature on the side scan, this may indicate that the anomaly is either an artifact (noise) or that the object is buried.

Geological features such as faults, grabens, dikes, intrusions and anticlines can be discerned, but may be very weak (fig. 10.7), and should be correlated with subbottom seismic reflection records.

The features that generate magnetic anomalies may be reduced to a simple geometric shape (sphere, rod, plate, etc.) on the basis of the anomaly signature and mathematical models (4, 6). It is also possible to calculate the depth of burial of a given object, such as a length of drillpipe, by assuming a model (in this case a vertical cylinder), by dissecting the signature of an anomaly and applying mathematical techniques involving the fall rate or signal decay away from the feature (6).

The final step in interpretation involves the plotting of the values of the anomalies, with respect to the survey navigation lines (see chap. 11). If cross lines (tie lines) have been run in the vicinity of an anomaly they should be checked to verify the presence of an anomaly. This process helps determine the nature of certain features; for instance, when a series of anomalies of equal value align, they are most likely due to a long linear object such as a cable or pipeline on the sea floor.

OPERATION AND CALIBRATION

Operational considerations to be made toward running a magnetometer during the course of a geohazards survey include:

The determination of the sensor towing distance behind the survey vessel that will not produce electromagnetic interference on the magnetometer recordings (fig. 10.6). In the instance of a survey conducted within a small nonferrous hull boat, with an outboard engine, it may be more convenient to place the magnetometer sensor ahead of the survey boat.

The distance off the sea floor at which the sensor is to be deployed. This height is critical toward the sensitivity of the magnetometer, as the closer the objects to be detected, the greater the amplitude that will be produced.

The calibration of the equipment for the particular geographical area should be in close agreement to the values shown in figure 10.2.

The fine tuning of the mangetometer system and selection of an appropriate sampling rate provides the effective resolution. Thus coverage will be restricted to the sampling interval along the ground, which may be computed at the rate of 50 cm/sec for every knot of the survey vessel. For example, a sampling rate of 1 sec (fairly common) will produce a sampling interval of 2.5 m, for a vessel traveling at a speed of five knots.

The desired resolution must be matched to the size and spacing of the objects sought. By the same token, the spacing of the survey grid must also accommodate the necessary resolution for resolving the sizes of features it is desired to detect.

The conduct of a magnetometer search survey for pipeline, ship, flight recorder, wellheads, tools, chain and anchors, etc., requires establishing whether the object is ferromagnetic (iron or steel only) or not, and the probability of detection by a magnetometer. If a search is to be performed in the vicinity of large metallic structures, such as a bridge or drilling platform, detection may be preempted by the large magnetic field generated by the structure.

The establishment of an appropriate survey grid spacing that would provide the necessary areal coverage must also be made. A square grid is best for providing both a complete and objective coverage of an area.

The background noise level must also be taken into account, in that it must be lower than the expected signature of the searched object. Thus if the noise level is averaging $+/-$ 3 γ, it would be quite difficult if not impossible to detect an anomaly of only 3 γ.

It is also feasible to use the magnetometer to relocate a position by marking such positions with a small permanent magnet.

The strong magnetic field generated by AC and DC powerlines may also be used to advantage in locating electrical conductor cables, by ensuring that these are acting as conductors during the search.

Magnetometers provide a valuable addition to the conduct of high-resolution geohazard surveys when attempting to either find ferromagnetic objects, or distinguish between nonferrous and ferrous targets.

Eleven
MARINE GEOENGINEERING HAZARDS

CASE STUDIES AND INTERPRETATION

Now that we have covered the tools of the survey trade, let us look at a few examples of their application through the interpretation and presentation of the survey results. A series of four case studies is presented in order to acquaint the reader and prospective user with the interpretive process, while a few valuable refrences are cited (1,2,3, and 4).

GEOENGINEERING HAZARDS

The term *hazards* as used herein is rather strong, and somewhat of a misnomer, but applies nevertheless to features (geologic and other) on and beneath the sea floor, which could pose a threat to either the stability or operation of an engineering structure in this environment. The use of the term constraint is perhaps more appropriate, in that most so-called hazards once identified can be dealt with or circumvented by proper design or choice of location.

The end product of a geoengineering survey should be a map of the geologic structure, geotechnical properties, and any anomalous features at a prospective offshore construction site (e.g., drilling), or along a route (e.g., pipeline or cable). Such maps should depict not only the bathymetry, but also any and all pertinent features such as outcrops, lateral changes in geotechnical properties, surface features indicative of currents and sediment composition (e.g., acoustically hard versus soft bottom), location of debris, unidentified magnetometer anomalies and side scan sonar targets, faults, and bright spots (amplitude anomalies).

Accompanying the maps should be a brief, yet concise, verbal report of pertinent information gleaned from the interpretation of the HRG survey data. The report should also include such items as: dates of survey, participating personnel (in case questions arise during the course of construction or operations), equipment models employed and their scale and settings, sensor deployment positions relative to the navigational receiver antenna, and if available, a brief summary (with references to published literature) of the local geology.

Some of the above items may be inserted as appendices, including the analysis of core samples, anomaly locations, and identification. Also a useful addition to the report is the inclusion of reproductions of pertinent survey records illustrating any features of concern.

The report's summary and conclusions should be to the point, and oriented above all to the engineers who will be responsible for the design, emplacement, construction, and eventual operation of the structure.

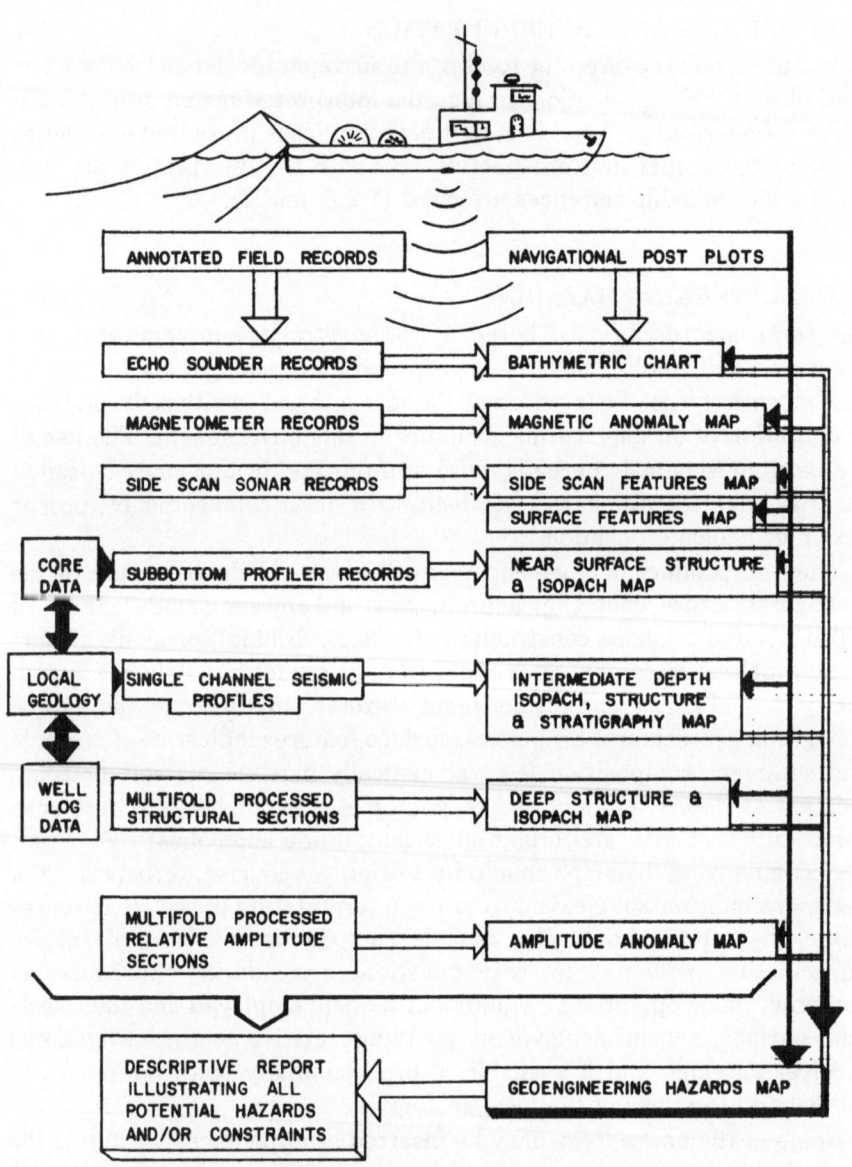

FIGURE 11.1 Flow-diagram for HRG survey data interpretation sequence. (By the author.)

INTERPRETATION SEQUENCE

An organized data flow sequence, during the course of an interpretation, assures both completeness and that all features will be accounted for.

The navigation postplot of the survey must be available along with the original annotated field records or preferably copies of the records. Copies of subsurface reflection profiles allow the pushing of colors, to assist in the correlation of reflectors, and pertinent markings for the identification of such features as faults, and prominent reflecting horizons, while producing a record of the interpretation.

The survey postplot map must contain appropriate geographic coordinates (latitude and longitude, as well as local grid), scale, geographic projection, grid orientation, lease block or concession boundaries, and if present, navigational aids and fairways.

The interpretation should proceed in a logical sequence, from the surficial sea-floor features on to the deeper stratigraphy, as shown by the flow chart of figure 11.1. Thus the bathymetry should first be completed and mapped, followed by the magnetometer data. Once mapped, the magnetometer data can then be doublechecked for cross-identification of anomalies with the side scan sonar data. Similarly, upon reviewing the side scan sonar records, these should be cross-checked against both the echo sounder results and magnetometer records for any anomalous features.

The subbottom profiler data should reveal a clear picture of the near-surface structure and stratigraphy. Faults should be doublechecked, after being located from deeper seismic profiles, with the subbottom and echo sounder records to verify their shallowest expression as an indication of most recent activity. If available, shallow core data should be projected onto the subbottom records for the identification and correlation of the seismic reflectors.

Single channel and multifold processed sections allow the mapping of faults and other structural features, as well as amplitude anomalies indicative of shallow gas accumulations.

Separate rough maps should be made for each of these sensors, for later compilation into the final display maps (e.g., surface features and hazards map). The report should include as few maps as possible. It is frequently possible to summarize the results on a single geoengineering hazards map, which would include bathymetric contours, magnetometer anomalies and side scan sonar targets, faults (with fault planes mapped to greatest depth detected), surface sediment variations (e.g., relict channels), and location, extent, and depth of amplitude anomalies (bright spots).

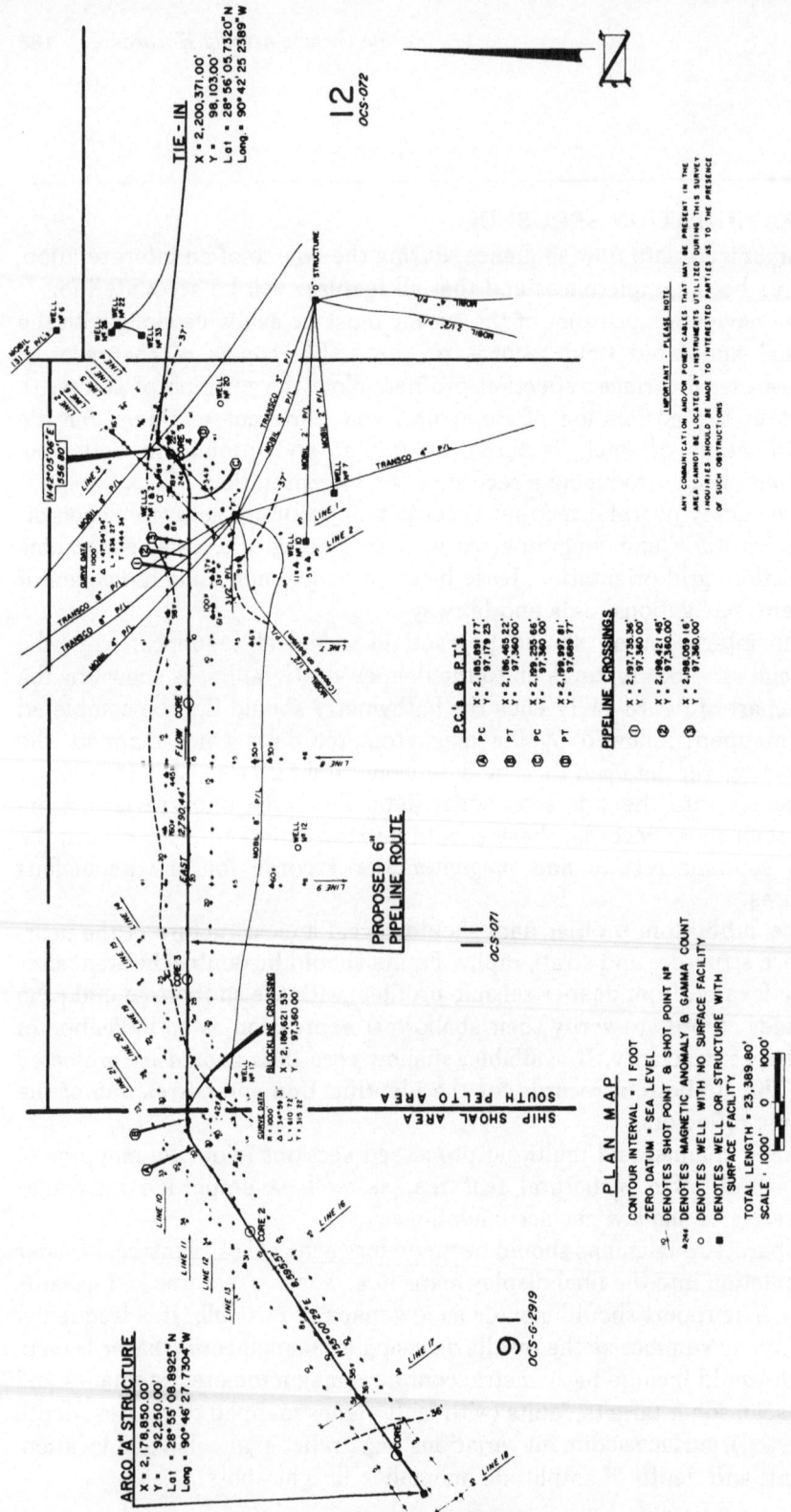

FIGURE 11.2 Navigational postplot and bathymetry "plan-view" of HRG pipeline route survey. Gulf of Mexico. (Courtesy of ARCO Oil & Gas Company, South Texas District, Houston, Tex.)

GEOENGINEERING SURVEY OBJECTIVES

Needless to say, it is quite pertinent that the bathymetry around an area of proposed construction be known, and that the sea floor be clear of debris (dumped ordnance, lost construction items, pipelines and communication cables, and sunken vessels), as such items could hamper the deployment of anchors, or setting of jack-up drilling rig legs.

The subbottom structure and stratigraphy should reveal the presence of faults and lateral variations in sediment composition and geotechnical properties, which could hamper the insertion of structural support piles into the sea floor, trenching of the sea floor for the burial of cables and pipelines, or initial drilling operations within the uppermost 100 m. Knowledge of the shallow stratigraphy is necessary to anticipate the reaction of an emplaced structure during and after installation, such as the setting of a jack-up platform leg into softer sediments of a shallow buried channel.

Case Study 1: Pipeline Pre-Lay Survey

The following geoengineering study was performed prior to the construction of a 15 cm diameter (6 in) natural gas pipeline between a production structure (ARCO "A" Structure), and an existing pipeline (fig. 11.2). The survey was conducted in the northern Gulf of Mexico.

The HRG survey was performed with the following sensors: Navigation (electronic ranging system), side scan sonar, magnetometer, and a subbottom profiler doubling as an echo sounder. No deeper penetrating seismic reflection system was used, in that the shallow objectives of a few meters, required for pipe-laying operations (trenching and anchoring), were adequately covered by the above sensors.

In addition to the HRG systems, a three meter gravity corer was used to obtain samples of near-surface sediments for geotechnical analyses.

The survey track lines consisted in three parallel lines, with the center coincident with the proposed pipeline route (fig. 11.2). A number of tie lines, perpendicular to the main lines, were also obtained for bathymetric and subsurface correlation purposes. The survey grid provided coverage of the sea floor to several hundred meters to either side of the construction route, in order to assure sea-floor clearance for the laying of anchors during pipeline construction operations.

The mapping of the interpreted survey results are displayed in the plan and profile views of figures 11.2 and 11.3. The interpretation has revealed the presence of a number of pipelines, oil and gas production structures, unidentified magnetic anomalies (fig. 11.4) and a near-surface growth fault (fig. 11.5). The bathymetry is contoured (dashed lines) at an interval

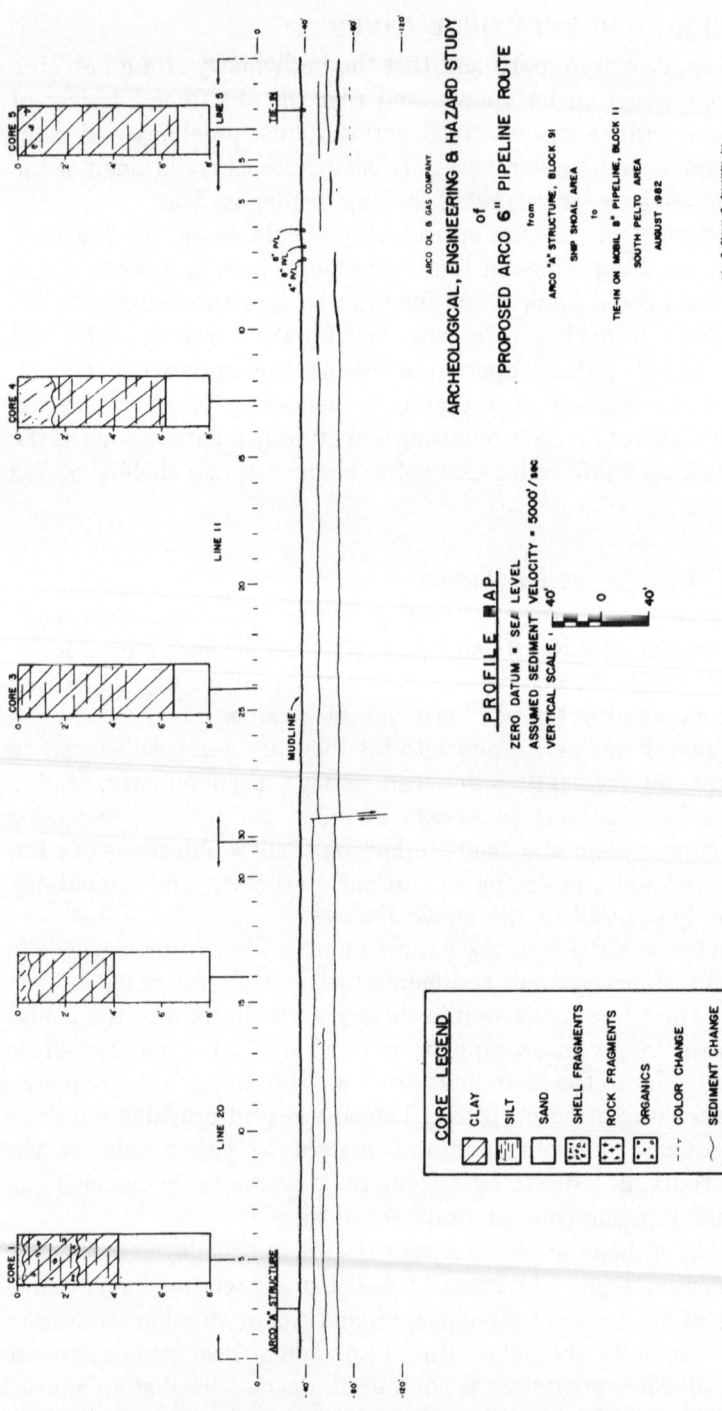

FIGURE 11.3 Profile "map" along proposed pipeline route, based upon interpretation of HRG survey records and geoengineering core data, Gulf of Mexico. (Courtesy of ARCO Oil & Gas Company, South Texas District, Houston, Tex.)

Line	Shot Point	Gamma Count	Signature Width (in ft)
1	1.9	9	50
2	4.0	11	50
2	4.6	5	25
5	4.6	24	35
5	5.9	34	125
6	2.0	16	75
6	4.7	8	50
6	5.5	18	150
7	5.6	6	25
8	5.9	50	50
8	6.5	30	Spike
9	6.6	60	75
10	2.8	22	125
10	6.4	8	150
10	9.9	98	75
10	16.6	445	50
10	18.9	760	75
12	6.3	57	25
12	6.6	13	25
12	7.7	100	50
12	8.2	61	75
15	3.1	42	35
17	3.1	24	50
19	18.6	4	50
20	12	8	100
20	8.5	315	100
20	7.8	5	50
20	7.7	36	50
21	15.0	31	50

FIGURE 11.4 Tabulation of magnetic anomalies detected by HRG geoengineering survey along proposed pipeline route, Gulf of Mexico. (Courtesy of ARCO Oil & Gas Company, South Texas District, Houston, Tex.)

of one foot with respect to an arbitrary sea level datum. Survey lines are identified by number and direction, while the navigation fix marks (shotpoints) are appropriately marked along the tracks for correlation with the survey records. Lease block boundaries are outlined as well as the local geographic grid (x's and y's). Additional annotations are listed regarding construction details for the proposed pipeline, and the precise location of pipeline crossings.

Data presented on the profile map (fig. 11.3) include the sea-floor elevations (bathymetry), and a line drawing of the shallow stratigraphy, based upon the subbottom seismic profiles, and the analysis of core

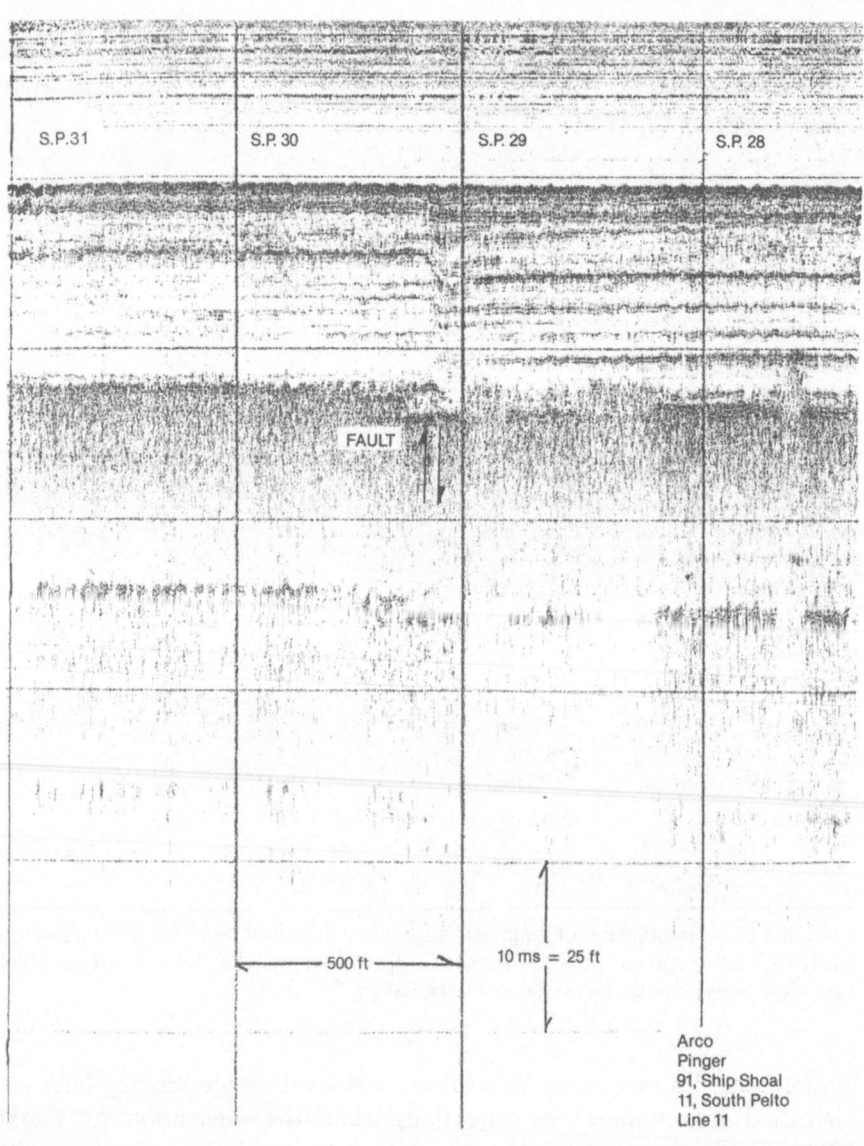

S.P. 31   S.P. 30   S.P. 29   S.P. 28

FAULT

← 500 ft →   10 ms = 25 ft

Arco
Pinger
91, Ship Shoal
11, South Pelto
Line 11

FIGURE 11.5 HRG subbottom ("pinger") profile (Line 11) along pipeline route (see fig. 11.2, for location), Gulf of Mexico. (Courtesy of ARCO Oil & Gas Company, South Texas District, Houston, Tex.)

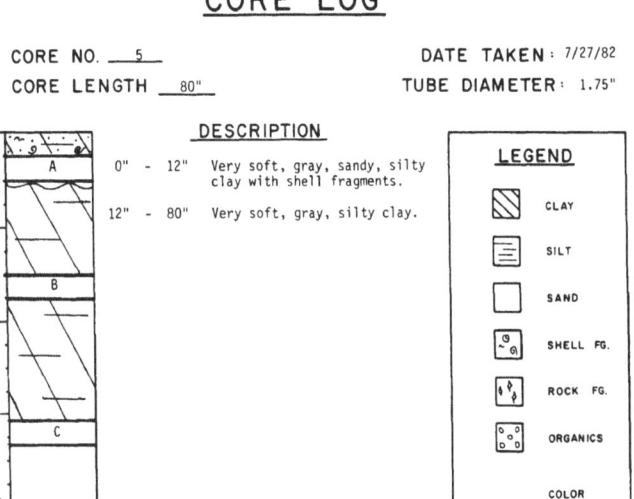

CORE LOG

CORE NO. ___5___   DATE TAKEN: 7/27/82
CORE LENGTH ___80"___   TUBE DIAMETER: 1.75"

DESCRIPTION

0" - 12" Very soft, gray, sandy, silty clay with shell fragments.

12" - 80" Very soft, gray, silty clay.

LEGEND

- CLAY
- SILT
- SAND
- SHELL FG.
- ROCK FG.
- ORGANICS
- COLOR CHANGE
- SEDIMENT CHANGE

SAMPLES

A
Dw. 101.3	Dd 61.4
Mc. 65.1	Uc. too soft
Vs. 107	Msg. 1.62

CLASSIFICATION: Very soft, gray, sandy, silty clay with shell fragments.

B
Dw. 95.5	Dd 54.7
Mc. 74.6	Uc. too soft
Vs. 96	Msg. 1.53

CLASSIFICATION: Very soft, gray, silty clay.

C
Dw. 90.5	Dd 45.8
Mc 97.4	Uc. too soft
Vs 72	Msg. 1.45

CLASSIFICATION: Very soft, gray, silty clay.

FIGURE 11.6 Log of geotechnical properties and sediment description for core samples obtained along pipeline route, Gulf of Mexico. (Courtesy of ARCO Oil & Gas Company, South Texas District, Houston, Tex.)

samples. Besides the descriptive sediment classification of the cores, geotechnical properties including shear strength (*Vs*), water content (*Mc*), dry weight (*Dw*), and specific gravity (*Msg*) were obtained, and presented separately for each core in the appendix as illustrated in figure 11.6.

In summarizing the results for this survey, it was recommended that the magnetic anomaly of 315 γ, at shotpoint 8.5, line 20 (figs. 11.4 and 11.7), be avoided or investigated by divers prior to pipeline construction.

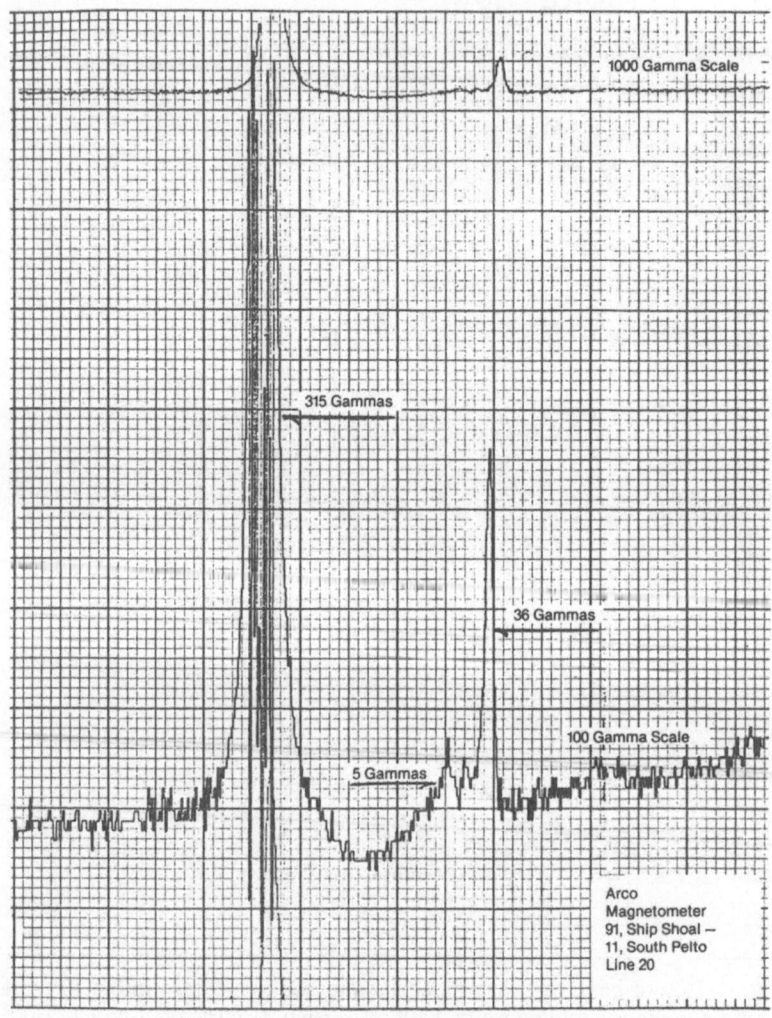

FIGURE 11.7 Magnetometer record (Line 20) showing anomalies along proposed pipeline route (see map of fig. 11.2 for locations), Gulf of Mexico. (Courtesy of ARCO Oil & Gas Company, South Texas District, Houston, Tex.)

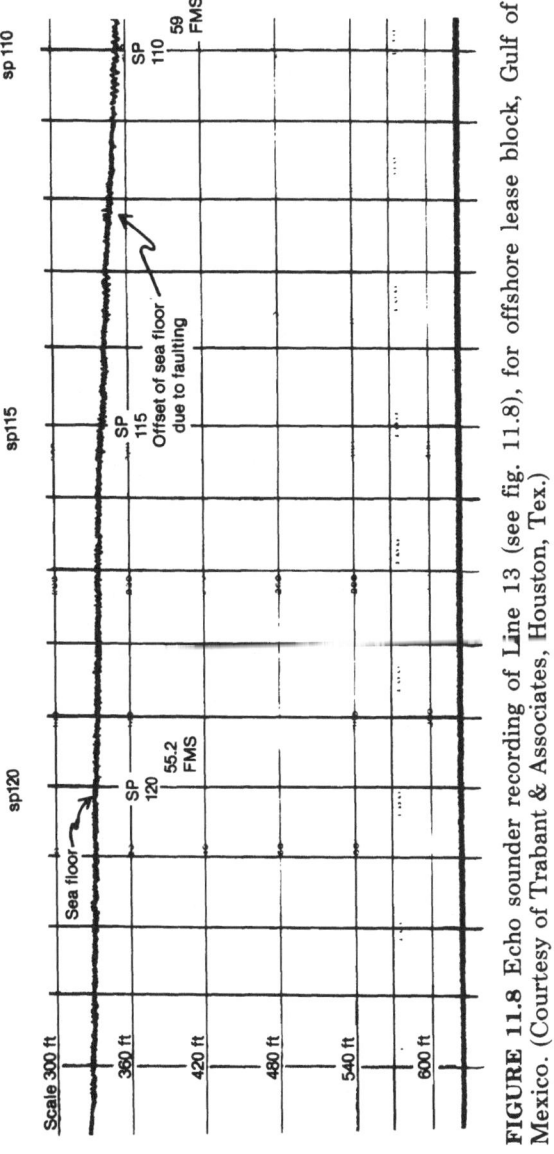

FIGURE 11.8 Echo sounder recording of Line 13 (see fig. 11.8), for offshore lease block, Gulf of Mexico. (Courtesy of Trabant & Associates, Houston, Tex.)

Evaluation of this anomaly during the survey could have been performed by running an additional survey tie line or two, while verifying the anomaly would have pinpointed the total amplitude and location of the magnetic anomaly. The amplitude of this particular anomaly should lead one to check for a possible abandoned drill site and casing, in view of the dense drilling activity in the area.

Water depths along the proposed route ranged from 10 to 12 m (35 to 37 ft), and the surface sedimentary strata, within which the pipeline was to be buried, consisted in very soft to soft silty clays. The sediments were considered to provide a stable foundation for the pipeline.

The fault was found to reach within three feet of the sea floor (fig. 11.5), with no displacement at the sea floor (mud line), and therefore not considered a hazard for the anticipated twenty year lifetime of the pipeline.

The report also noted that no channel cut and fill features were present, and no problems were anticipated with regard to soil conditions.

During the installation of pipelines, the sea floor is usually jetted (soft clays or loose sands) or trenched (firm clays or compacted sands) depending on the soil (sediment) type. Thus, in this survey the report could state that no problems are expected in trenching. In the case that slight changes in sea-floor properties are detected, different trenching methods may have been required along portions of the pipeline route.

Case Study 2: A Structural Puzzle

The following case involves a geoengineering survey conducted near the edge of the continental shelf in the northern Gulf of Mexico. The area encompassed by the survey included a narrow salt dome spine that produced a horst structure. The uplifting of the salt also triggered mass movements of surface sediments. The navigation postplot is plotted on the respective maps.

Equipment deployed for this survey consisted of an echo sounder, subbottom profiler, magnetometer and minisleeve exploder source for a twelve trace multifold CDP recording. However, the CDP data were not processed for the interpretation, which therefore had to rely upon the near-trace analog monitor recordings.

Water depths ranged from 96 to a maximum of 116 m (312 to 381 ft) (figs. 11.8 and 11.9), with an average gradient for the sea floor of 0.2% above the 103 m (340 ft) contour, below which the slope increased to 0.53% (fig. 11.8, at sp 115).

The subbottom profiler and near-trace records of figures 11.10 and 11.11 are oriented north-south across the horst, as mapped on the shallow

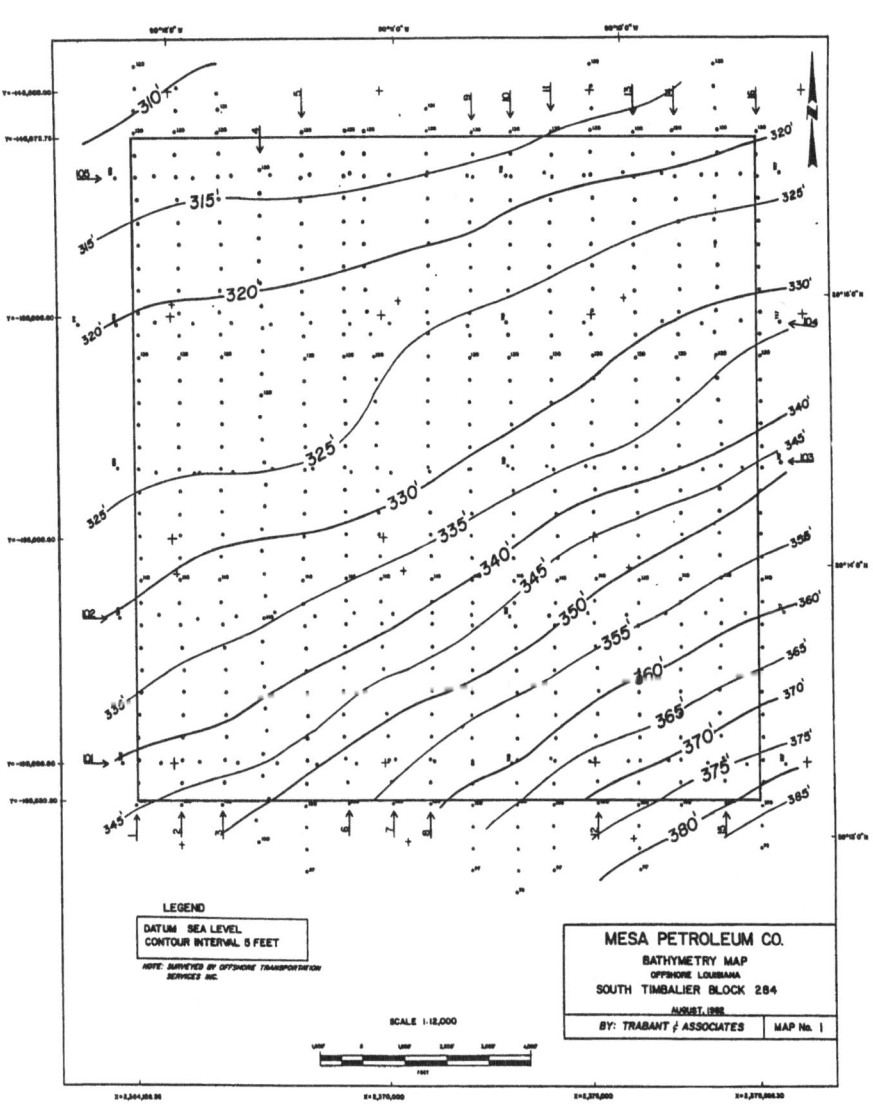

FIGURE 11.9 Bathymetric contour map generated from HRG geoengineering survey, for offshore lease block, Gulf of Mexico. (Courtesy of Mesa Petroleum Co., Houston, Tex.)

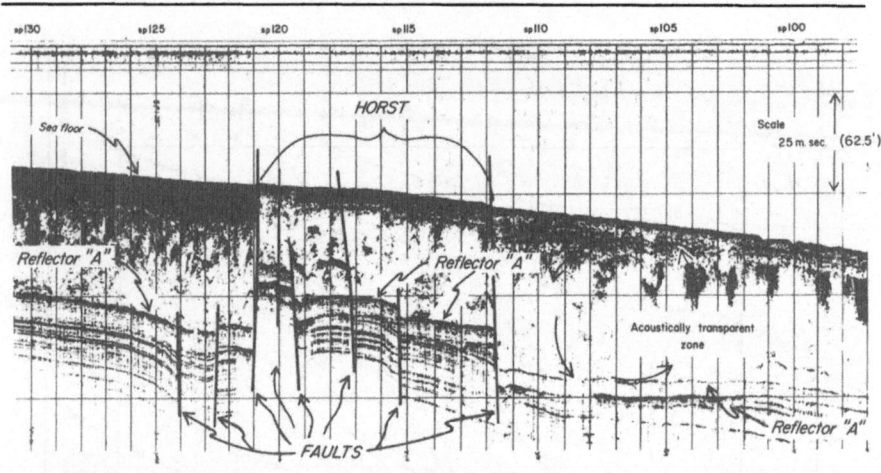

FIGURE 11.10 Subbottom profiler recording (Line 13, fig. 11.8), with interpreted faults, for offshore lease block, Gulf of Mexico. (Courtesy of Trabant & Associates, Houston, Tex.)

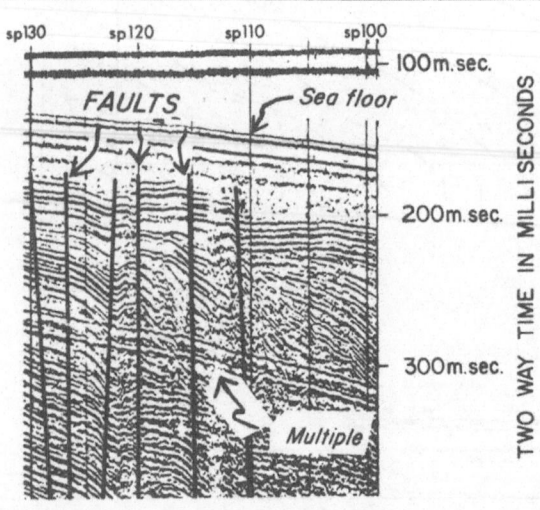

FIGURE 11.11 Single channel HRG analog minisleeve recording (Line 14, see fig. 11.8 for location), Gulf of Mexico. (Courtesy of Trabant & Associates, Houston, Tex.)

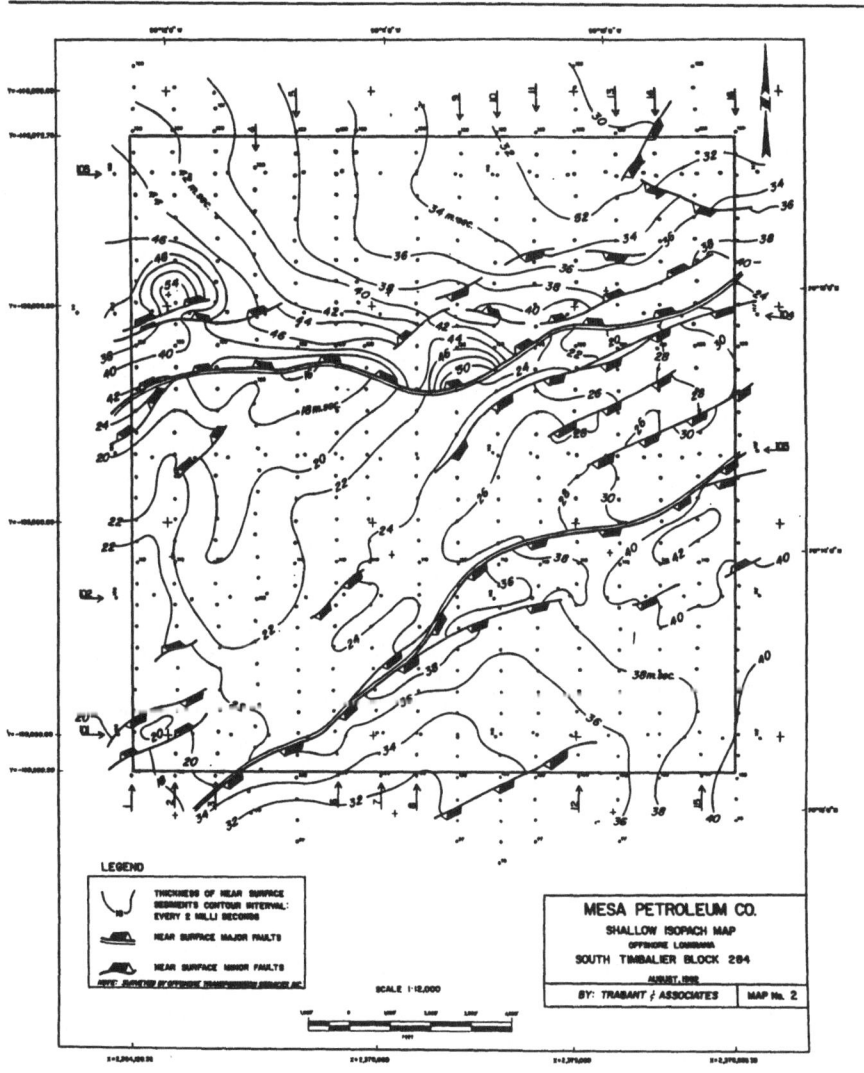

FIGURE 11.12 Shallow isopach map, derived from subbottom profiler data (fig. 11.11), from offshore lease block, Gulf of Mexico. (Courtesy of Mesa Petroleum Co., Houston, Tex.)

structural isopach of figure 11.12. The echo sounder data displayed surface offsets along the major fault systems (fig. 11.8), indicating recent movement of the faults.

The jumbled incoherent nature of the uppermost 30 m of sediment (fig. 11.10) is attributed to mass sediment movements, triggered by the uplifting of the salt spine, and the presence of biogenic gas within these sediments. The combination of vertical movement, and resulting mass sediment displacement, made interpretation tricky at first. The judicious pushing of colors along a few of the most prominent reflectors (e.g., reflector A, fig. 11.10), and demarcation of the complex faults, however, finally allowed the complete tying of the reflectors on all lines within the entire survey area. The shallow isopach map of figure 11.12 depicts both the near-surface sediment thickness (in msec), and the shallow structure with respect to the sea floor.

Recommendations, based upon the interpretation of these data, were that extreme caution be followed in the selection of potential drilling sites, in view of the possible instability of the uppermost 30 m of surficial sediments. The hazards map (fig. 11.13) showed a small stable area within the northwest corner of the survey area as well as the detected fault planes, bright spots, and magnetometer anomalies.

Case Study 3: Anatomy of a Blowout

The following HRG geoengineering survey was performed as the result of a blowout and the loss of a drilling rig in the northern Gulf of Mexico. The drilling site had not been subjected to a HRG survey prior to drilling operations. The following events chronicle the steps that led to the loss:

After the installation of a jack-up rig at the site, operations proceeded with the drilling of a pilot hole to a depth of about 340 m (fig. 11.14a).

An 80 cm casing pipe was then installed and cemented to a depth of 150 m (fig. 11.14b), an an annular blowout preventer (large valve) put in place at the sea floor.

The well then kicked and began flowing (water and drilling mud) to the surface. At this stage the preventer valve was activated, but the flow of fluids continued up and around the outside of the casing pipe (fig. 11.14c).

The drilling rig began to founder and all personnel were evacuated. The rig then proceeded to capsize and disappeared below the surface during the night. The following day gas was observed boiling on the sea surface over an area 200 m in diameter, and up to 2 m in height (fig. 11.14d).

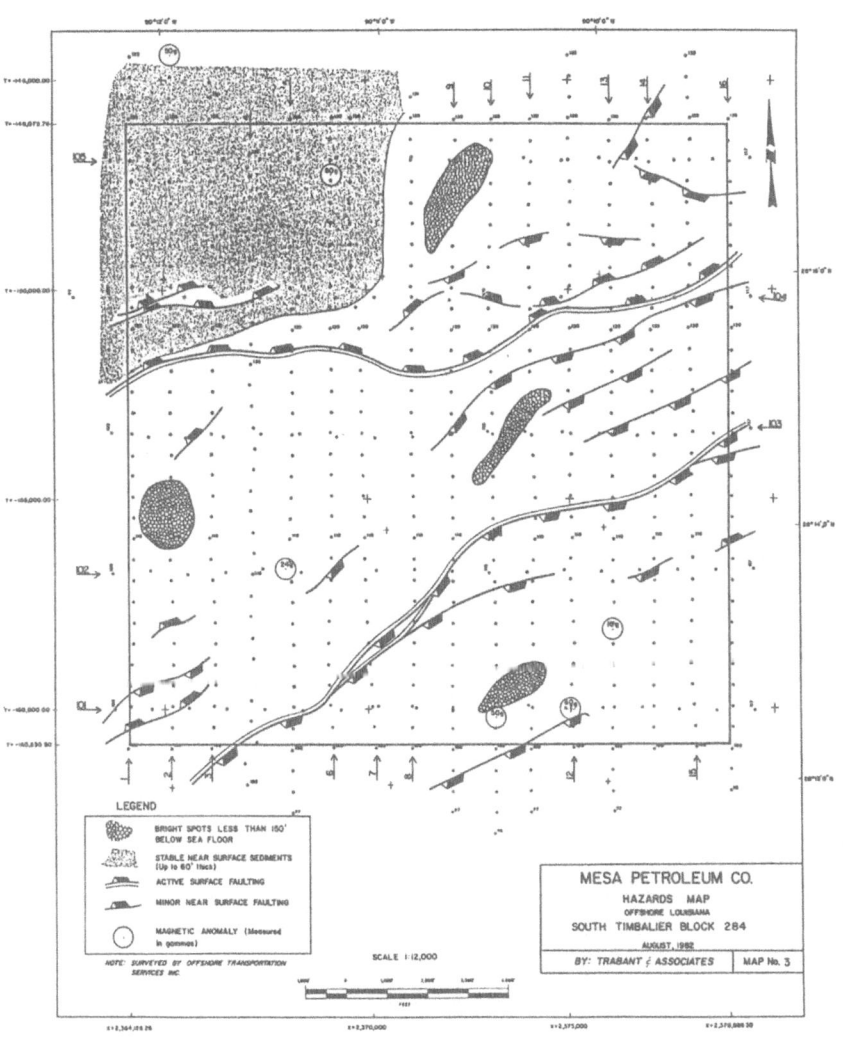

FIGURE 11.13 Hazards map, showing geohazards to the emplacement of drilling structures. (Courtesy of Mesa Petroleum Co., Houston, Tex.)

The gas flow eventually bridged, and ceased flowing after six days, with the total loss of the drilling rig, no injuries to personnel, and no pollution.

The cause of this typical blowout was the presence of a shallow gas sand at a depth of 330 m, as illustrated by both the structural and relative

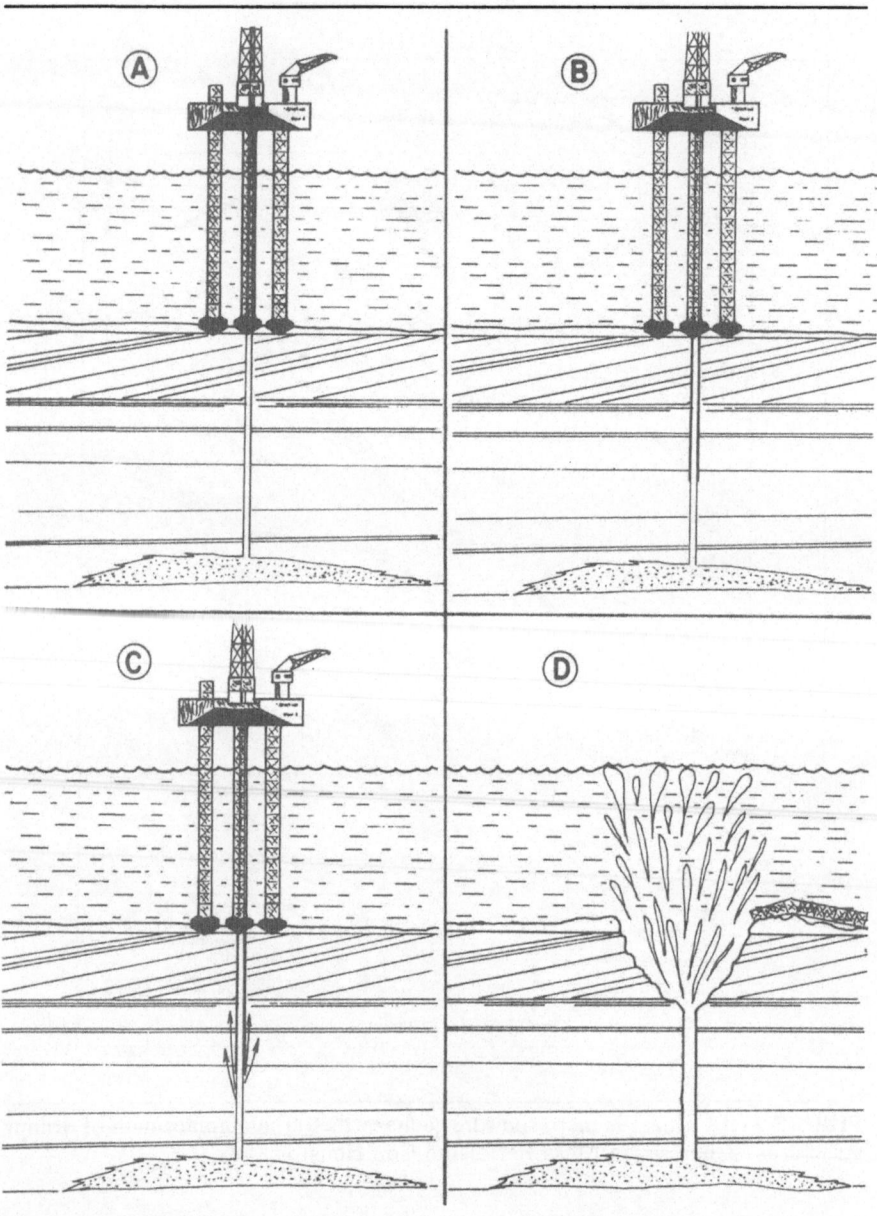

FIGURE 11.14 Illustration of sequence of events leading to a blowout: (A) Jack-up drilling rig "spuds" initial pilot hole to a depth of 340 meters, (B) casing (80-cm diameter) set and cemented to depth of 220 meters, blowout preventer installed at sea floor, (C) overpressured gas flows out of gas sand formation, within open hole and around casing to sea floor, (D) large flow of gas entrains sediments, removing foundation of drilling rig, which collapses. (Drawing by the author.)

amplitude processed multifold seismic sections of figures 11.15 and 11.16. The amplitude anomaly was very broad, and extended over an area of several kilometers. While the amplitude anomaly was apparent on the processed multifold structural seismic sections, it could not be discerned on the single channel analog monitor record (fig. 11.17).

Had these processed CDP survey data been available, and properly interpreted, prior to drilling operations, recommendations should have been made to the effect that an amplitude anomaly indicative of shallow gas accumulation was present at a subsurface depth of 330 m. Furthermore, upon establishing the drilling program (targeted depth, casing sizes and depths, and drilling mud weights), the drilling engineer should have taken such a recommendation into consideration. The pilot hole could then have been drilled short of the amplitude anomaly at 330 m, and the casing pipe firmly cemented in place. Following this operation, diverters and blowout preventers could have been installed prior to perforating the anomaly, with the use of heavier-than-normal drilling mud fluids. Once the shallow gas sand was perforated, casing should have been installed and cemented prior to proceeding with drilling operations.

These latter steps would have most likely assured the successful drilling of the well on the basis of the HRG geoengineering survey data, and interpretation as shown above.

A number of after-the-fact HRG geoengineering surveys over blowout sites and jack-up drilling rig loss sites, caused by unexpected leg penetration, have shown that the large majority of these losses could have been prevented by the proper interpretation of quality HRG survey data.

Case Study 4: Deltaic Mud Flows

The rapid deposition of fine-grained deltaic sediments, off the mouth of major rivers, produces thick wedges of unstable sediments. In the following case, off the Mississippi River Delta, sediments have been deposited at rates averaging up to 30 cm per year for the past several centuries. This rapid accumulation has not permitted the dewatering, or normal consolidation process, to take place, and resulted in the accumulation of high water content, soupy sediments. When distributed by the passage of storm waves, the low-strength sediments move downslope under the influence of gravity. Mud flows of this nature occur on the prodelta areas of several major rivers, where petroleum exploration and production operations are active.

Thus when contemplating drilling operations, or the installation of a structure within such an environment, it is critical that a thorough

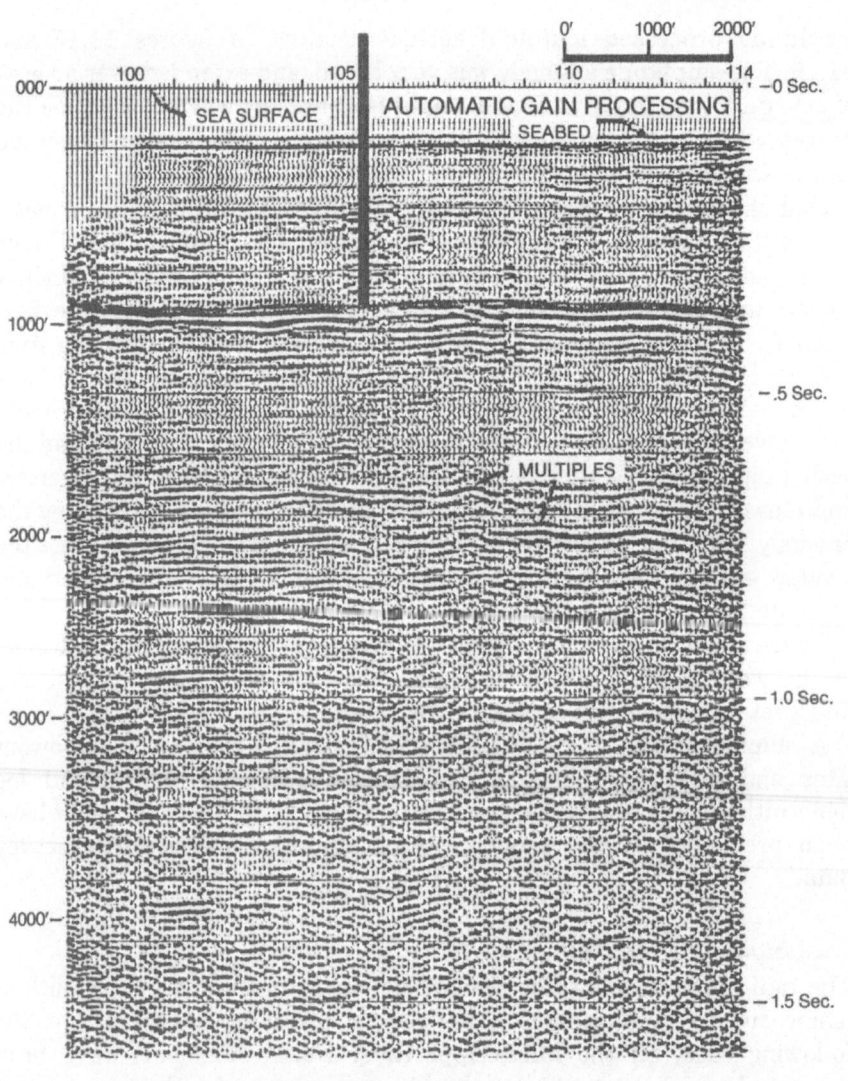

FIGURE 11.15 Processed multifold (12-channel) sparker seismic reflection record section over blowout site. Note strong reflector (at .36 seconds or depth of 940 feet), which contained overpressured gas. (Courtesy of Trabant and Associates, Houston, Tex.)

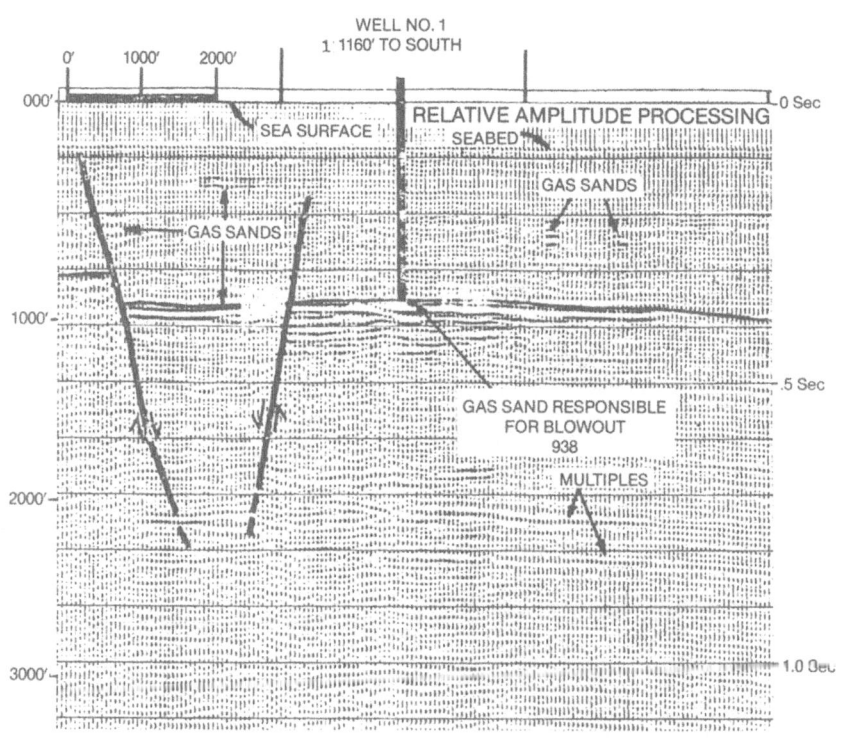

FIGURE 11.16 Same seismic section as fig. 11.15, with "relative amplitude processing," revealing blowout causing amplitude anomaly from shallow gas-sand (at 0.36 sec., or depth of 940 feet). (Courtesy of Trabant & Associates, Houston, Tex.)

mapping of the sea floor be carried out in order to ascertain that the rig, or structure emplaced, will not be dragged downslope by mass sediment movement.

It was toward this end that the following data were collected, and that an area considerably larger than the targeted drilling site was surveyed, in order to map in detail the mud flow features in the vicinity (particularly upslope) of the proposed drill site location.

The HRG sensors deployed consisted of a subbottom profiler, magnetometer, boomer, and side scan sonar. The latter was a digital unit that removed the water column and corrected for slant range, in order to assemble a photomosaic picture of the sea floor.

Past studies (5) of this particular area had revealed that mass sediment failures occur within well-defined channels or gullies, and circular areas

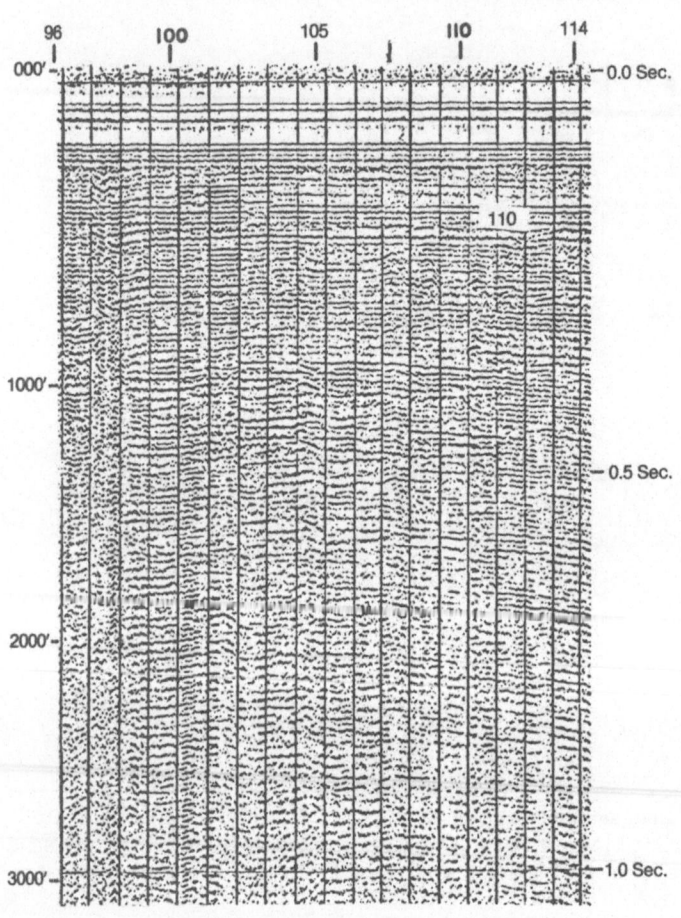

FIGURE 11.17 Single channel analog section of seismic sections of figures 11.15 and 11.16. Note absence of amplitude anomaly where blowout occurred (shotpoint 113). (Courtesy of Trabant and Associates, Houston, Tex.)

known as collapse depressions. Past experience with deeper penetrating HRG seismic systems had produced no data, as a result of the high biogenic gas content of the recently deposited sediments. Thus no powerful HRG system was deployed for this particular survey.

The results of the interpretation were presented as a series of maps of the survey track (fig. 11.18), bathymetry (fig. 11.19), surface features (figs. 11.20 and 11.21), and an isopach of the near-surface sediments.

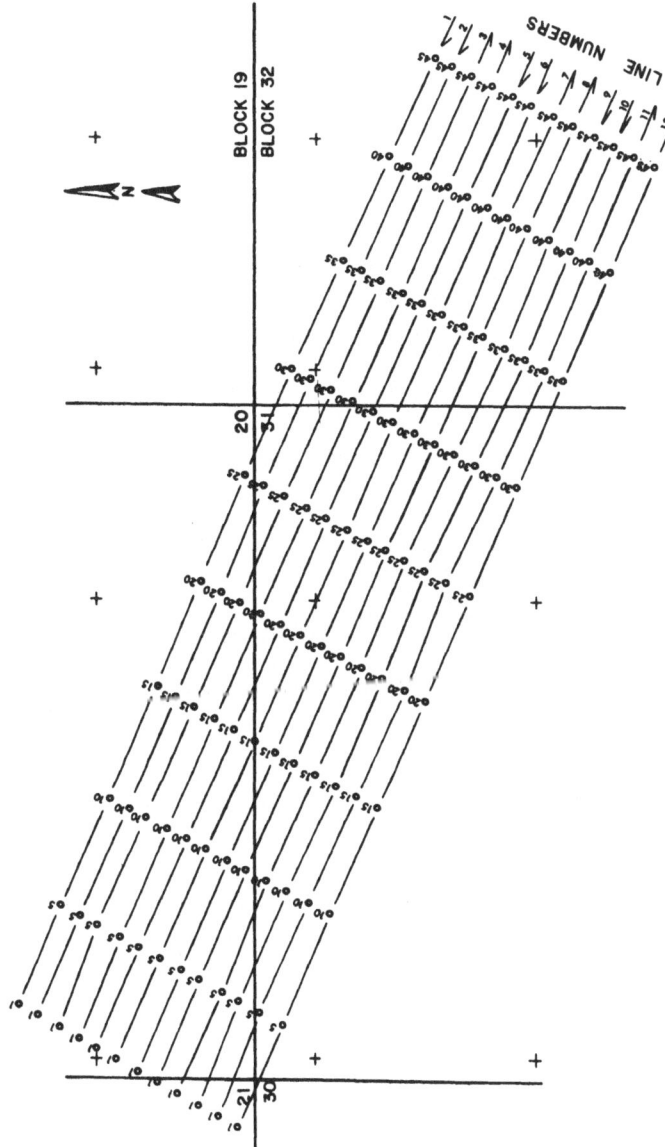

FIGURE 11.18 Navigational postplot map, for geoengineering hazards survey off Mississippi River Delta, Gulf of Mexico. (Courtesy of Hunt Oil Company, Dallas, Tex.)

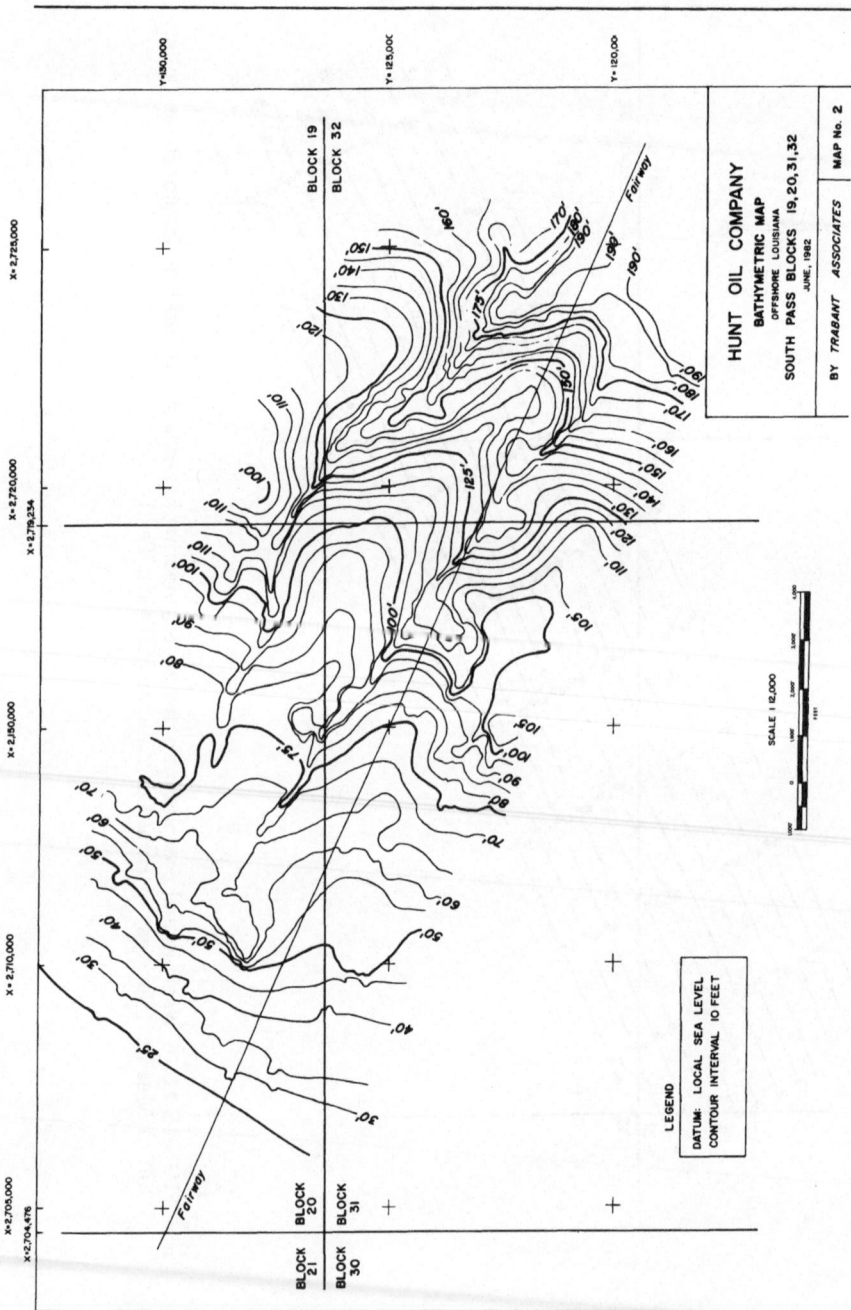

FIGURE 11.19 Bathymetric map produced from HRG geoengineering survey data, Mississippi River Delta, Gulf of Mexico. Note indentation of contours associated with mud-flow gullies. (Courtesy of Hunt Oil Company, Dallas, Tex.)

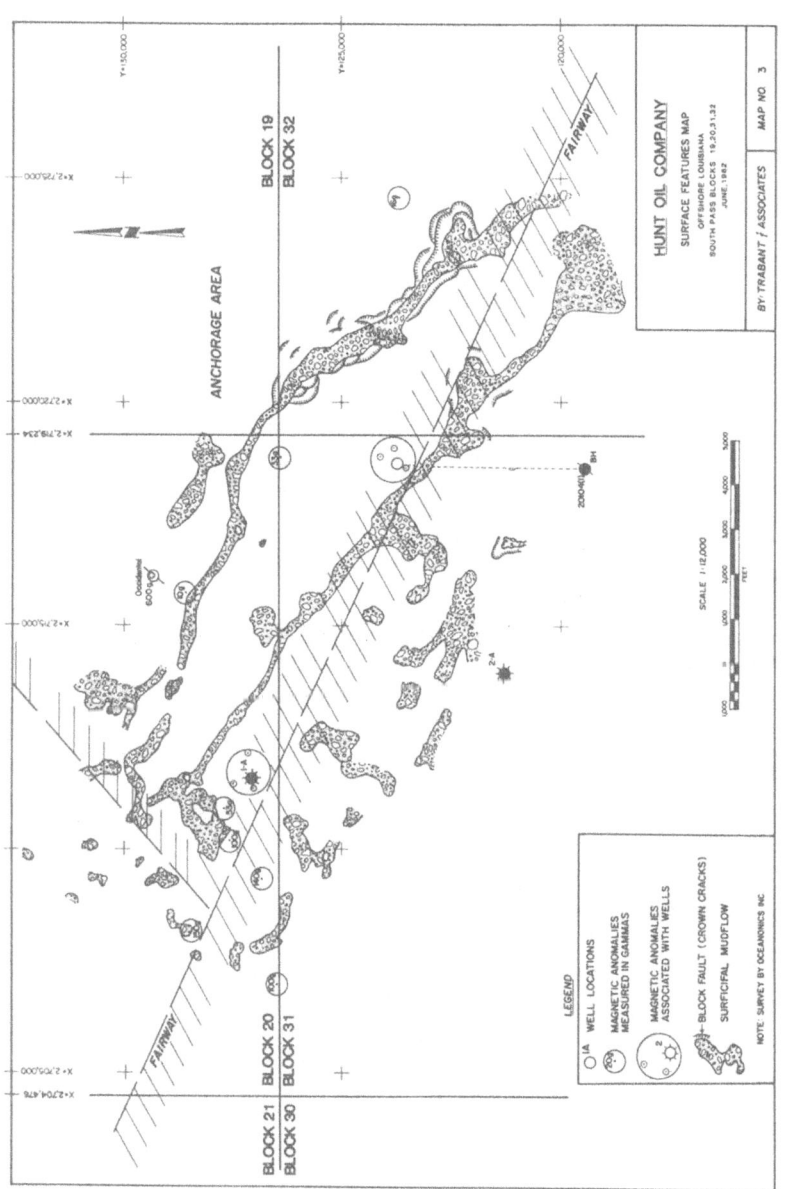

FIGURE 11.20 Surface features map, based upon interpretation of HRG geoengineering hazards survey data, Mississippi River Delta, Gulf of Mexico. (Courtesy of Hunt Oil Company, Dallas, Tex.)

The bathymetry revealed an uneven topography as a result of the presence of two elongate downslope channels, as shown in figure 11.19.

Review of the side scan sonar, and subbottom profiler records revealed the presence of two active mud flow gullies (figs. 11.20 and 11.21) associated with the bathymetric channels (figs. 11.22 and 11.23).

The features were mapped and presented as illustrated in figure 11.20, along with the detected magnetometer anomalies (mostly the result of past drilling activity in the area).

Based upon past experience, the driller was given a certain safety margin, or distance from the borders of the mud flow features, where a drilling rig could be placed. In this particular case it was imperative that precise navigation be adhered to, for the safety of the rig while on location, since the drilling site was located on the margin of a mud flow, and anchors would have to be located outside of the features, lest movement pull the anchors (and drilling rig) downslope.

The effects of the presence of acoustically opaque zones, caused by the presence of biogenic gas, are illustrated by the boomer profile of figure 11.24.

In conclusion, great care and very detailed HRG surveys are necessary within deltaic areas. The inability to obtain HRG data from subsurface depths leaves the door open to a large degree of uncertainty as to the presence of such shallow geohazards as amplitude anomalies and faults within the uppermost 1000 m. It should also be noted that mud flow features are not restricted to deltaic areas as their presence is also common along the continental slope as exploration proceeds into deeper waters.

SEISMIC INTERPRETATION

The interpretation of marine HRG recordings requires a knowledge of marine geology, and the past and changing nature of geologic processes within marine and near-shore environments. The following comments pertain to a few specific items and geologic processes with which the interpreter of HRG marine data should be familiar.

Sea Level Fluctuations

An important controlling factor of the shallow geology and geomorphology of the continental shelf areas is the fact that over the past many thousands of years, sea level has risen, lowered, and stood still as a result of the waxing and waning of large continental ice sheets.

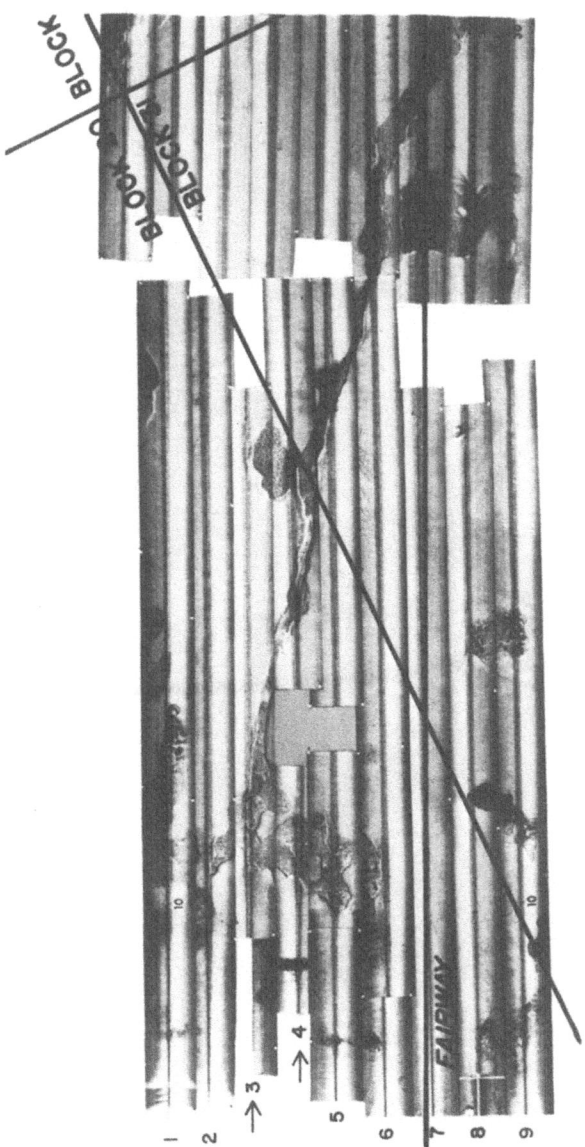

FIGURE 11.21 Photographic mosaic, assembled from nine adjacent HRG side scan sonar recordings, depicting mud-flow gullies and collapse depressions, Mississippi River Delta, Gulf of Mexico. (Courtesy of Hunt Oil Company, Dallas, Tex.)

208

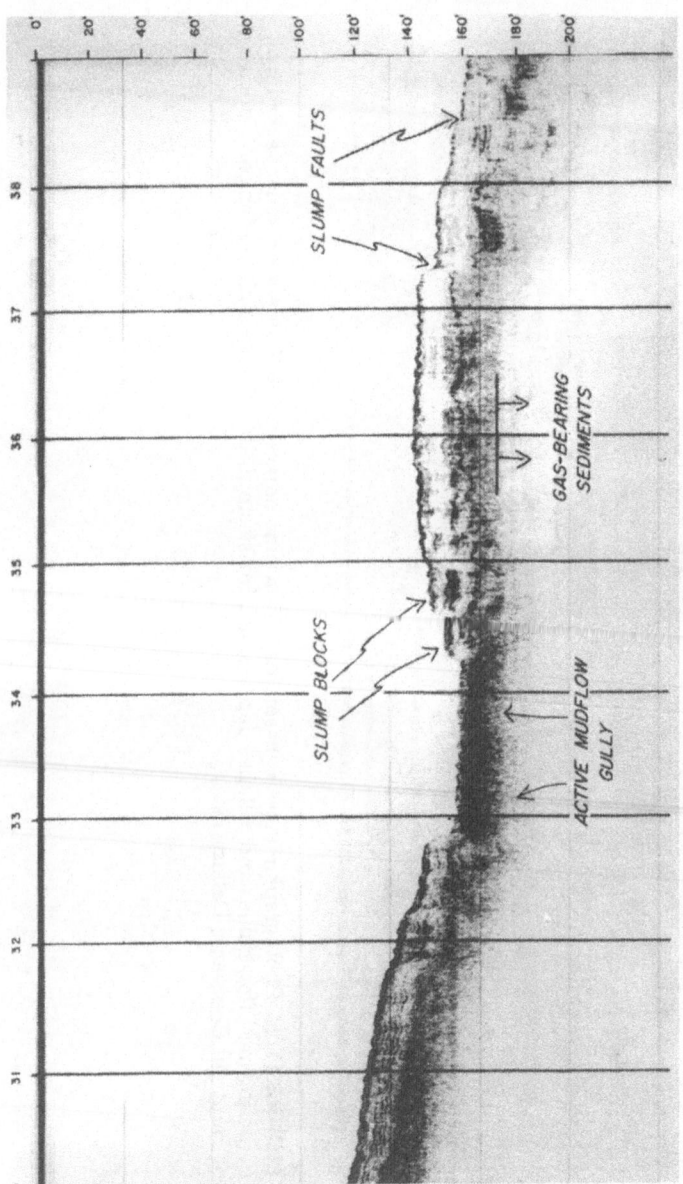

FIGURE 11.22 Subbottom (7.0 kHz) HRG survey record (Line 2 see fig. 11.18 for location), over mud-flow gully, Mississippi River Delta, Gulf of Mexico. Note that subbottom records were also used for obtaining water depths. (Courtesy of Hunt Oil Company, Dallas, Tex.)

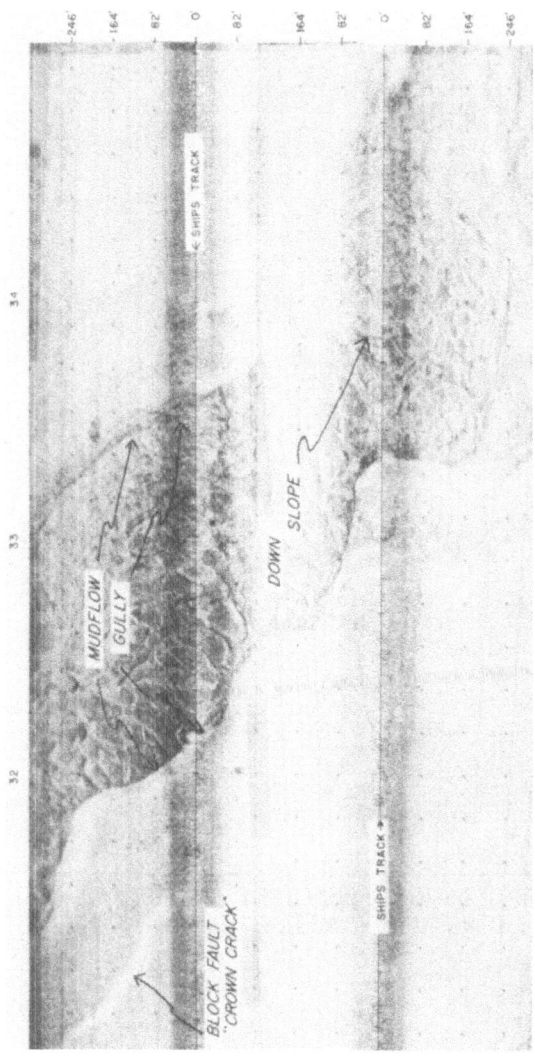

FIGURE 11.23 Pair of adjacent side scan sonar records (Lines 2 and 3, see location in fig. 11.18), showing details of a mud-flow gully, Mississippi River Delta, Gulf of Mexico. Note "blocky", surface texture resulting from downslope flowage of surficial sediments. (Courtesy of Hunt Oil Company, Dallas, Tex.)

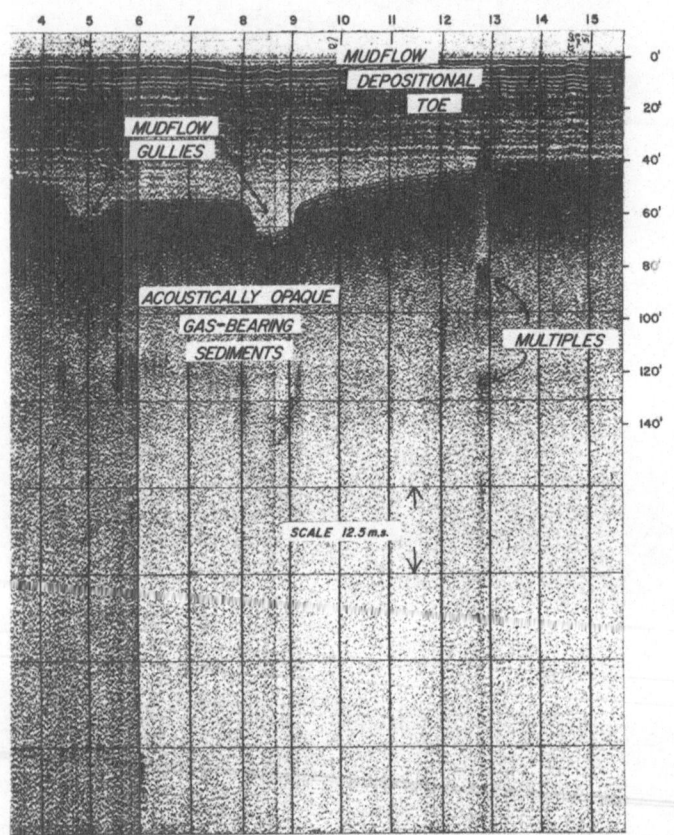

FIGURE 11.24 Boomer recording (Line 4, see fig. 11.18 for location) over mudflow features, Mississippi River Delta, Gulf of Mexico. (Courtesy of Hunt Oil Company, Dallas, Tex.)

These processes have resulted in the exposure of large portions of the continental shelf (areas now submerged) to the effects of desiccation, erosion, and glaciation, processes that presently take place along coastal areas.

Features such as rivers and lakes from periods of low stands in sea level are referred to as relict, and can frequently be identified on HRG subbottom profiles. Other features resulting from these processes include relict beaches and shorelines (including dunes), deposits of glacial till and boulders, gullies, canyons, and valleys and stiff soils formed by the desiccation of older marine clays.

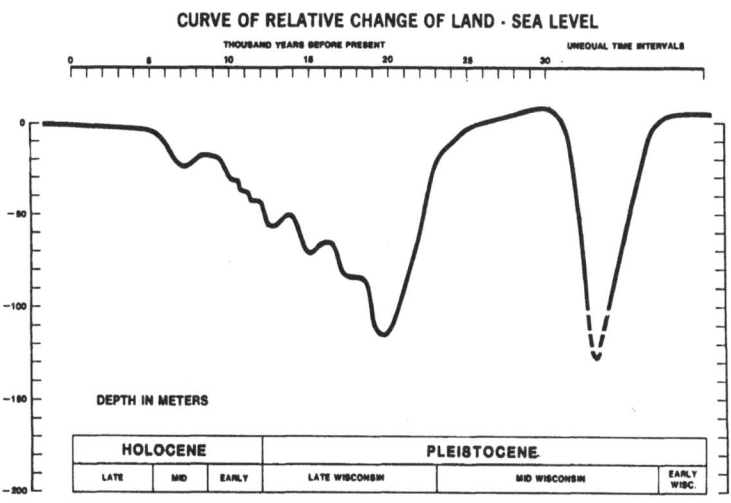

CURVE OF RELATIVE CHANGE OF LAND · SEA LEVEL

FIGURE 11.25 Curve depicting relative change of sea level versus time, for the northern Gulf of Mexico. (Courtesy of Coastal Environments, Inc., Baton Rouge, La.)

Depending upon local tectonism, sea level curves based upon radio-carbon and other dating techniques may be graphed as illustrated in figure 11.25, for the northern Gulf of Mexico (6) as an aid to interpretation. Variations in geotechnical properties of near-surface sediments are frequently a result of these eustatic or worldwide changes in sea level.

Faults may be identified on HRG seismic reflection profiles depending upon the fault type, angle, and the amount of displacement. The matching of reflectors across growth and wrench faults is more often a difficult exercise, and a careful process of tieing key reflectors within a survey area is necessary to assure the proper interpretation. Where active faults extend to the sea floor, they are frequently detectable on both echo sounder and side scan sonar records. The mapping of fault planes at depth (2000 m or more), on the other hand, requires the use of deeper penetrating seismic exploration data.

Bright spots or amplitude anomalies usually require relative amplitude processing of multifold HRG data for their detection. These features indicate the presence of strong acoustic reflections, which may or may not be the result of gas accumulations or even overpressured gas deposits, which could result in a blowout during drilling operations. Thus, care must be exercised in their identification. It should also be noted that processed multifold HRG records are not fail-safe. In some cases, HRG records obtained where blowouts have been caused by shallow gas accumulations did not indicate the presence of amplitude anomalies.

Twelve
SURVEY ORGANIZATION, NAVIGATION, AND FUTURE DEVELOPMENTS

SURVEY ORGANIZATION

Given a specific site or area requiring a geoengineering evaluation, certain items need to be considered prior to initiation of field surveys. These include the specific survey objectives, time frame, location, local geology, anticipated geoengineering hazards, weather, logistics, optimum equipment, and selection of qualified personnel. The sequence of major steps in this process are outlined in figure 12.1.

Survey Objectives

The anticipated geotechnical properties of the sea floor and type of engineering structure to be installed are paramount in the formulation of a proper geoengineering survey. Typical structures include pipelines, cables, anchored semisubmersible drilling rigs, jack-up drilling rigs, or construction of support island and platforms. The geotechnical properties at the sea floor, and underlying strata, will establish the foundation type and suitability for a particular structure.

General guidelines for the selection of HRG equipment, given the anticipated hazards and survey objectives (type of structure) are presented in Table 12.1.

Time Frame

The time required in carrying out the presurvey organization, the survey proper, and postsurvey interpretation need to be realistically assessed. Some of the factors that control this may include drilling rig operational constraints and scheduling, or timing of an appropriate fair weather window or survey season.

Location

Also critical is the specific location requiring a survey. Working with an already well-founded vessel in one's own back yard, with ample logistic support, presents a much simpler operation than one halfway around the globe in poorly charted waters, far from support facilities.

Local Geology and Environment

Familiarity with the local geology from either past experience, or a detailed review of previous surveys and available literature, is an important step toward establishing minimum survey requirements and anticipated results. The range of water depths, sea-floor morphology, and geologic structure should serve as a basis for establishing the orientation and spacing of the survey grid.

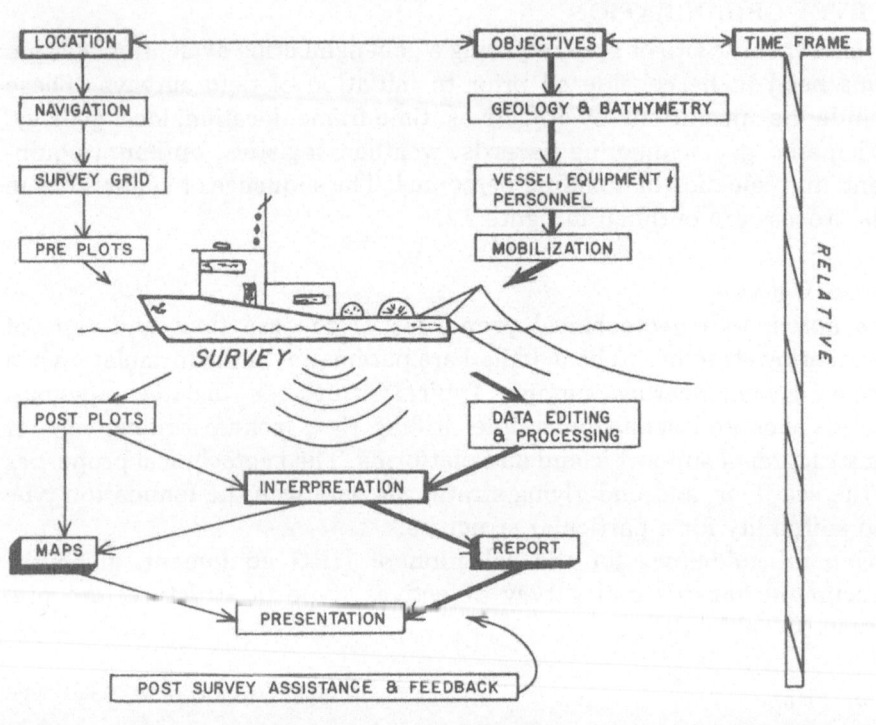

FIGURE 12.1 Flowchart of survey organization, operations, and interpretation sequences. (By the author.)

The anticipated geology should also establish the necessary high-resolution systems to satisfy coverage of both the sea floor and subbottom strata and meet the specific survey objectives. These factors will also determine the required type of survey vessel. If shallow waters are prevalent (near shore, bays, and reef areas), a shallow draft vessel will be required. On the other hand, if the site is within an area of expected rough seas such as the Arctic or North Sea, a large seagoing vessel may be required.

Optimum Equipment

The selection of appropriate HRG tools should be based on the local geology and specific survey objectives. Other factors to be considered are: sea-floor coring operations, multifold data acquisition, the need for long-range side scan sonar, or the deployment of a deep-tow system. If a deep-

TABLE 12.1 Equipment selection guide for given survey objectives.

SURVEY OBJECTIVES	GEOENGINEERING HAZARDS				
	SEA FLOOR OBSTACLES	RELICT CHANNELS	GEOTECHNICAL PROPERTIES	SHALLOW GAS	FAULTS
BOTTOM SUPPORTED RIG	X	X	.	X	X
ANCHORED DRILL – SHIP (BARGE)	X	X	.	X	X
PILE SUPPORTED PLATFORM	X	X	X	X	X
PIPELINE	X	X	.	.	X
SURVEY EQUIPMENT					
ECHO SOUNDER	X	.		.	.
SIDE SCAN SONAR	X	.		.	
MAGNETOMETER	X	.		.	
SUBBOTTOM PROFILER	X	X	.	.	X
BOOMER	.	X	.	.	X
DEEP - TOW	(For water depths greater than 200 meters)				
SINGLE CHANNEL ANALOG	.	X		.	X
MULTI – CHANNEL DIGITAL CDP	.	X	.	X	X

tow survey is to be performed, the survey vessel must accommodate not only the bulky towing equipment, but also be capable of maintaining very slow speeds (1–2 kts) for extended periods of time. The deployment of either STDs (salinity, temperature and depth) or velocimeters is highly recommended for deep water areas to obtain accurate acoustic velocities in the water column.

In addition to adequate spares, instruction manuals and repair equipment must also be carried or be available on short notice. The necessary power requirements and space for the equipment must also be considered in the choice of an appropriate survey vessel. Certain equipment requiring large electrical current, such as sparkers, need a separate power source so as not to interfere with other equipment power requirements. Automatic or manual event-marking electrics should be installed between all recorders and the navigator or navigation equipment.

Logistics

Clearances must be obtained from proper authorities prior to survey operations. These include survey permits, radio navigation, and radio communication licenses.

Reliable transportation of equipment and personnel to and from the survey site and vessel must also be available, as well as resupply schedules for fuel, water and groceries.

Since losses are expensive, the time required for the processing of documents and the mobilization of equipment, personnel, and survey vessel are very critical. For example, if one piece of equipment is tied up in customs or awaiting repairs, operations must be suspended or modified, at great expense to the operations. Good radio communications with shore-based personnel, and available transportation, is essential to avoid many of these problems.

Personnel

As space and external logistics are severely limited aboard ship, qualifications of each individual must be adequate to operate more than a single piece of equipment. Motivation, versatility, and good seamanship of the personnel are paramount to a successful survey. The number of personnel should also be sufficient for conducting operations. More would be needed for twenty-four-hour operations compared to those conducted during daylight hours only.

Good technical expertise and the ability to make the most with material at hand, can and has frequently saved the day for survey operations.

Necessary documentation (passports, visas, medical clearances, etc.) are also necessary for out-of-country operations.

Familiarity of the survey vessel's crew with the ship, local regulations, and the conduct of survey operations is a great asset.

Advance notification to proper authorities concerning fishing regulations may alleviate problems associated with the presence of fishing vessels, nets, and traps within the survey area.

Consideration of the above factors and experience will result in a smooth and timely operation, while omission of a single factor may result in time-consuming delays, or poor quality data acquisition, and inadequate coverage for the specific survey objectives.

MARINE NAVIGATION SYSTEMS

Navigation is the practice of directing the course of a survey vessel along a preestablished track or survey line, and the plotting of the vessel's precise geographic location on the water on a real-time basis (1).

Historically, the navigation of a vessel involved the observation of celestial bodies (sun, moon, planets and stars) when away from land, and bearings or ranges to prominent landmarks when in the vicinity of a coastline. Such observations result in the plotting of lines of position (LOPs), the intersection of which represents the position of the survey vessel.

Navigation at sea thus involves obtaining ranges and/or bearings on a continuous basis, to or from objects of known geographic position by means of visual, radio, or acoustical techniques. A minimum of two LOPs at appropriate angles of intersection are necessary for a good fix. The intersection of LOPs should be between 30° and 150° in order to minimize the area of uncertainty.

While visual survey methods are sometimes used for near-shore surveys, they have been largely replaced by radio positioning systems (RPS). Radio positioning provides accurate continuous position updates (fixes) with a minimum of personnel in both near-shore and far offshore survey areas.

There presently exist a large variety of navigational systems used in the conduct of HRG marine surveys. Most of these are of the RPS type listed in table 12.2.

The choice of a specific type of system will be predicated upon the survey location system availability, and desired positioning accuracy. If a

TABLE 12.2 Specifications for a number of radio-navigation systems.

System	Maximum Operating Range (Nautical Miles)	Accuracy (meters)	Transmit Frequency
OMEGA	Worldwide	3000–1500	10.2 kHz
LORAN-C	300–1500	500–50	100 kHz
NNSS Satellite	Worldwide	300–30	150/400 MHz
DECCA-SEAFIX (Mainchain)	300–600	500–30	100 kHz–1.6MHz
TORAN	100–400	250–30	1.4–1.8MHz
LORAC	100–400	250–30	1.6–1.8MHz
RAYDIST-N/RAC	100–200	250–20	1.6–2.4 MHz
DECCA HI-FIX	100–200	50–20	1.6–4.0 MHz
RAYDIST-DRS	100–200	50–20	1.6–4.0 MHz
ARGO	100–250	25–10	1.6–1.8 MHz
XR-SHORAN	75–150	200–25	230–400 MHz
SYLEDIS/MAXIRAN	60–250	20–5	200–450 MHz
CUBIC AUTOTAPE	15–50	5–2	2.9–3.1 gHz
ARTEMIS	15	25–2	9–10 gHz
MOTOROLA RPS/ MINI RANGER	20–60	12–2	3.5–10 gHz
DEL NORTE TRISPONDER	20–40	12–2	9–10 gHz

Source: Modified from Gay 1983.

system is already installed in a particular area, and it meets the accuracy requirements, it may be a more judicious choice than the installation of another system. If working in a remote area, consideration must also be given to logistics of weight and bulk of the system, as well as availability of preestablished geodetic markers.

The following descriptions are general in nature, in order to acquaint the reader with various systems currently in use. Details regarding individual navigation systems may be found in the literature or directly from the manufacturer.

Acoustic Systems

This technique consists in the deployment of a number of acoustic transponders, which upon receiving a signal from the survey vessel returns an acoustic pulse, to provide the distance as a function of the acoustic velocity of the signals through the water (2).

It is necessary to calibrate the position of the transponders, with the use of RPS or a SatNav system, and account for both the acoustic velocity and slant ranges, between the acoustic transponder units and survey vessel. Hence their maximum operating range is restricted and their use in the routine conduct of geoengineering surveys has been limited. However, such systems are of great value for the operation and navigation of submersibles and the positioning of deep-tow packages. Furthermore, they allow rapid deployment in remote areas, where shore-based facilities (antennas and transmitters) may not be reliable, or radio frequency allocation and licensing may be difficult to obtain.

Radio Frequency Systems

Radio positioning systems (RPS) consist in a number of transmission antennas (usually ashore, but also on islands, platforms, or buoys) that transmit pulsed radio signals; these signals are decoded by a receiver aboard the survey vessel (base station), to provide the location of the vessel with respect to either direct ranges to the antennas, or to hyperbolic lines called lanes, generated by the combination of several (master and slave) stations, as illustrated in figure 12.2. The distances using RPS are based upon the velocity through the atmosphere of the transmitted radio frequency signals.

The more common systems and associated parameters are listed in table 12.2. Each of these systems provide tradeoffs in terms of accuracy and maximum operating range, in a manner similar to the frequency and penetration dichotomy of HRG profiling systems.

The use of *satellite navigation* (Navy Navigational Satellite System, SatNav, or NNSS) has become common as an addition to the RPS systems, through the use of integrated navigation systems (INS). The NNSS provides highly accurate fixes, on the order of a few meters, in all parts of the globe at approximate hourly intervals (3). The frequency of fixes is a function of the latitude, satellite elevation, and number of satellites in operation. Because of their periodic passes, the NNSS system is not sufficient on a stand-alone basis, in that it does not provide continuous fixes, but rather serves as a precise method for the calibration of other systems.

The NNSS satellite navigation system transmits two frequencies and provides positions on the earth's surface by measuring the Doppler frequency shift as the satellite passes along a precisely known orbit with respect to time.

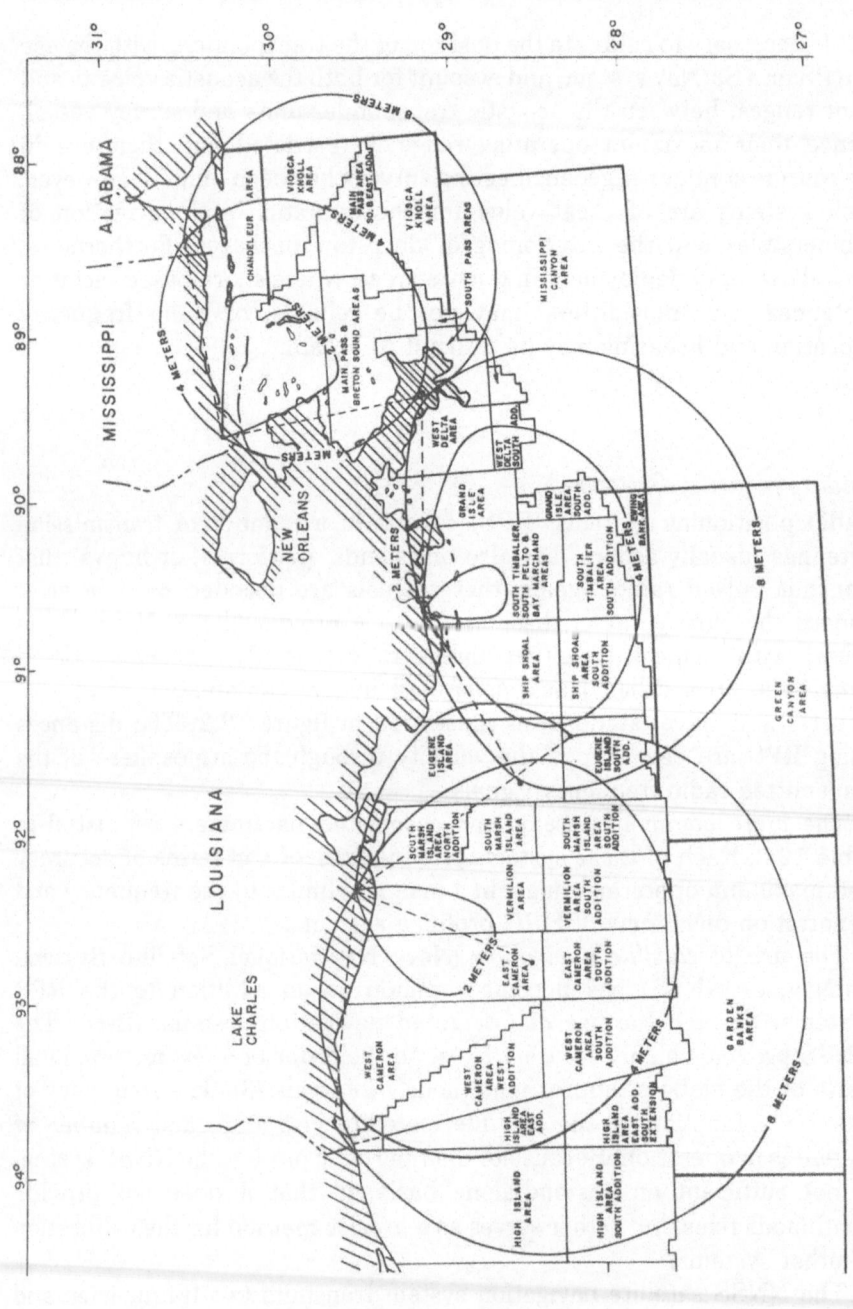

FIGURE 12.2 Areal coverage of several hyperbolic navigation networks, with accuracy contours (2, 4, and 8 meters), for the northern Gulf of Mexico. (Courtesy of Odom Offshore Surveys, Inc., Baton Rouge, La.)

Interfaced Navigation Systems (INS)

Besides recording and computing fixes from any of the RPS, acoustic, and the SatNav systems, INSs also serve to control the firing rate of seismic sources on the basis of distance traveled over ground (a necessity for the collection of CDP data). The systems are also used to record certain peripheral sensors, including ships' headings and speed logs, echo soundings, and magnetometer data.

SYSTEM SELECTION

Choice of navigational systems is dictated by a number of factors, ranging from system availability to desired survey precision.

The operational frequency selection may be critical, particularly if twenty-four-hour operations are desired, as a particular system may be subject to erratic positioning as a result of skywave and other atmospheric interference during nighttime hours. Similarly, the effects of strong sunspot activity may temporarily preclude precise navigation with many low-to-medium frequency RPS systems.

An ideal navigational system would provide (1):

continuous twenty-four-hour operation
worldwide coverage
all-weather capability
greater accuracies than presently available
standard coordinate system

While this may only represent a wish list at present (1984), it is expected that the Global Positioning System (NAVSTAR), due for launching in 1987, will provide all these factors, and bring a standardization to the numerous offshore navigation systems currently in use (3).

PRECISION AND REPEATABILITY

Accuracy is important along with repeatability for continuity of positions, particularly during postsurvey construction, and field installation operations. Thus if a particular hazard to the emplacement of a drilling rig were to be avoided by a small distance, the positioning of the rig must be extremely precise relative to that of the geoengineering survey data.

Precise knowledge where the survey data were collected is extremely important, as well as the on-track positions of the vessel for the common depth point technique, or in the deployment of sonobuoys.

A high degree of geodetic control of the land-based radio transmitter antenna location is required to provide the necessary accuracy of a given RPS navigation system.

CALIBRATION TECHNIQUES

Fundamental to the accuracy of a geoengineering survey is the calibration of the navigational system employed. This should be performed both before and after the survey, and daily if the survey is of lengthy duration.

Simple methods of system calibration involve the positioning of the survey vessel at a precisely known location (e.g., platform leg), or the crossing of baselines or their extensions, for range-range systems (fig. 12.3). The calibration should be carried out as close to the survey area as possible, and always doublechecked upon completion of the survey. Alternate techniques include cross-check with other precision systems, and/or obtaining 3 or more LOP solutions.

FUTURE DEVELOPMENTS

While relatively little research and development is carried out by the HRG survey industry, spinoffs from the petroleum industry, positioning companies, equipment manufacturers, and universities produce developments of value to the conduct of geoengineering surveys.

The electronic evolution of the past decade, which resulted in the mass production of inexpensive microchips, and fiber optic technology, are slowly (1984) seeing applications with time-proven software to reduce the size and weight of equipment, while improving data quality. Appreciable improvements in signal-to-noise ratios were accomplished with the conversion of seismic reflection data from analog to digital, and the advent of mass data storage methods and computer processing. The following techniques and equipment are discussed with respect to their general trend for future development.

Radio/Satellite

Communication linkage is currently available for the transmission via satellite of digital data. This capability could accommodate the direct transmission of raw data, directly from a survey vessel to shore-based installations for processing and interpretation. Such radio linkage could greatly speed up the normal turnaround time between survey completion and report delivery, particularly where large distances are involved.

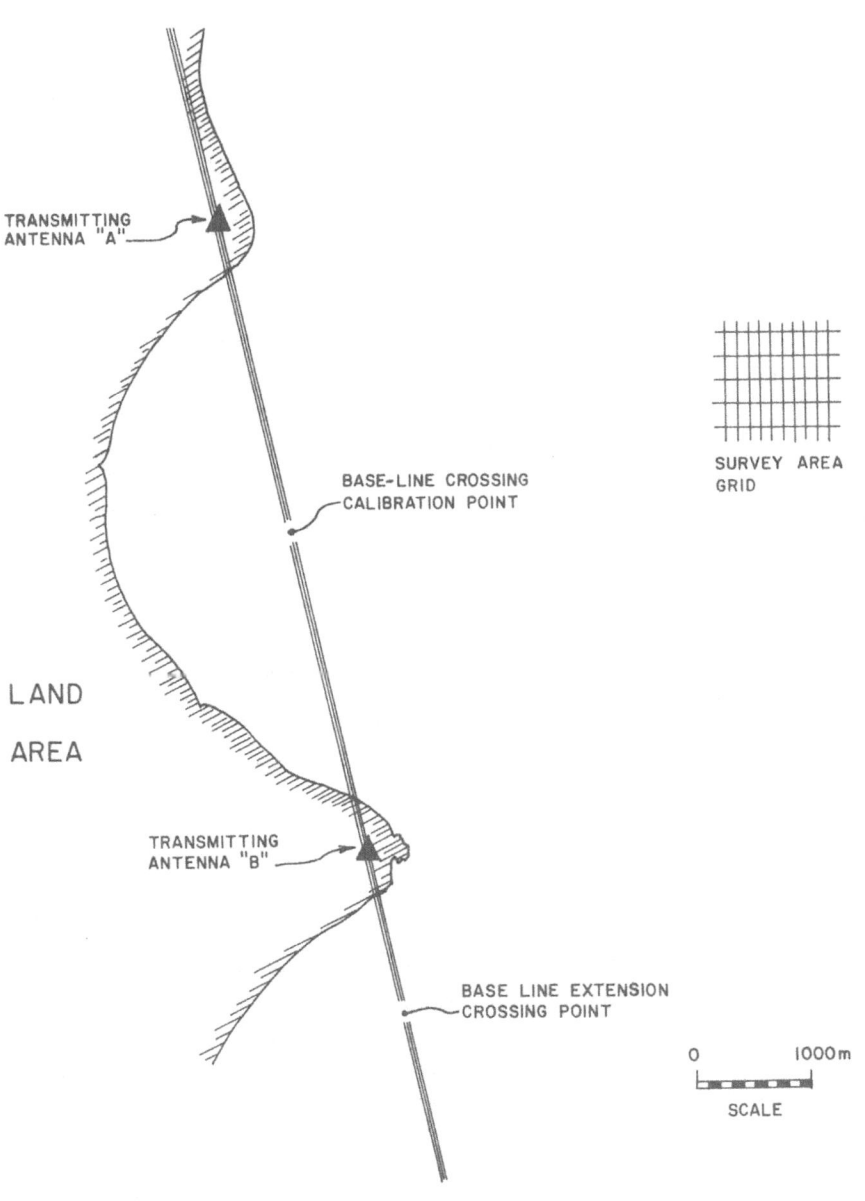

FIGURE 12.3 Calibration technique of range-range navigation system, by the crossing of baselines to verify distance between transmitting stations (A and B). (By the author.)

NAVSTAR

The expected launching in 1987 of the NAVSTAR, Global Positioning System (GPS) satellites, should provide the high-resolution industry with an optimum continuous navigation system of great precision (on the order of one meter) (3). The system should also eliminate the installation of costly and bulky transmitting equipment and antennas, which are now sometimes located at poorly controlled geodetic positions.

Single Channel Digital

The reduction in size of electronic microprocessors and mass storage devices allow data processing in real time on all single channel systems, including side scan sonars and subbottom profilers. While many systems presently available perform certain functions, such as the removal of the water column and correction of slant ranges for side scan sonar, better signal processing software should further enhance data acquisition compared to the previously all-analog systems.

CDP

The current trend within exploration geophysics, which has usually been a forerunner of the HRG industry, has been toward more phones, lighter digital/fiber optic cables, and real-time processing. It is thus expected that the high-resolution industry will eventually follow suit with yet further improvements in the single-to-noise ratio, resolution, and greater penetration of subsea strata.

CHIRP

Progress toward the development of multifrequency acoustic transmission, which may then be selectively filtered and collapsed (deconvolved) to produce optimum resolution, penetration, and signal-to-noise ratios relative to the monofrequency systems of the present, has been slow. Such systems as the CHIRP Sonar are expected to be available for routine HRG data acquisition within the near future.

Deep Tow

With petroleum exploration heading toward deeper water targets, the HRG surveys will have to accommodate these deeper environments for the eventual construction of production structures. It is also to be expected that the deployment of deep-tow multisensor packages will become routine in greater water depths.

Penetrometers

The use of sea-floor penetrometers, fired at great velocity, to provide a measure of the in-situ physical properties, as a function of the deceleration, should eventually reduce the presently required number of borings and core samples, easing the cost of obtaining sea-floor geotechnical properties and allowing rapid measurements during the course of a survey. While work has been in progress for over a decade toward refining such devices (4), penetrometers may become operational sometime in the future. Their use in deep waters will be most useful due to the difficulties of obtaining geoengineering borings in this environment.

Electroresistivity

The ability of electroresistivity measurements to provide data (formation factors) on subbottom strata, during the course of a survey by dragging appropriately spaced electrodes along the sea floor behind a survey vessel, should provide yet another tool for the routine investigation of engineering sites. The operation is similar to downhole logging operations, except performed in the horizontal at the sea floor (5).

The above trends in equipment development should lead to an overall reduction in the bulk of present-day HRG equipment, while providing better quality and more quantitative data, on the nature of the sea floor with respect to its performance as a foundation for the emplacement of engineering structures. Similarly, it is expected that new HRG tools will be developed for new environments of operation, such as ice or the abyssal depths of the ocean.

Finally, it should be noted that standardization of terminology between engineers, geologists, and geophysicists would be a most progressive step. It is hoped that the terms used herein, and presented in the glossary, might provide a step in that direction.

APPENDIX

TABLE OF USEFUL CONVERSIONS TO THE SYSTEME INTERNATIONALE (SI) D'UNITES

To convert from	To	Multiply by
AREA		
acre	meter^2	4.046 856 E + 03
foot^2	meter^2	9.290 304 E − 02
MAGNETISM		
gamma	telsa (T)	1.000 000 E − 09
FORCE		
kilogram-force	newton (N)	9.806 650 E + 00
kip	newton (N)	4.448 222 E + 03
LENGTH		
fathom	meter (m)	1.828 800 E + 00
foot	meter (m)	3.048 000 E − 01
League (Nautical)	meter (m)	5.556 000 E + 03
micron	meter (m)	1.000 000 E − 06
mile (International Nautical)	meter (m)	1.852 000 E + 03
mile (U.S. Statute)	meter (m)	1.609 344 E + 03
yard	meter (m)	9.144 000 E − 01
MASS		
pound-mass (1bm avoirdupois)	kilogram (kg)	4.535 924 E − 01
ton (long, 2240 1bm)	kilogram (kg)	1.016 047 E + 03
ton (short, 2000 1bm)	kilogram (kg)	9.071 847 E + 02
PRESSURE or STRESS		
bar	pascal (Pa)	1.000 000 E + 05
dyne/centimeter^2	pascal (Pa)	1.000 000 E − 01
kilogram-force/meter^2	pascal (Pa)	9.806 650 E + 00
millibar	pascal (Pa)	1.000 000 E − 02
pound-force/foot^2 (PSF)	pascal (Pa)	4.788 026 E + 01
TEMPERATURE		
degree Celsius	kelvin (K)	$t_K = t_C + 273.15$
degree Fahrenheit	kelvin (K)	$t_K = (t_F + 459.67)/1.8$
degree Fahrenheit	degree Celsius	$t_C = (t_F - 32)/1.8$

To convert from	To	Multiply by
VELOCITY (Includes Speed)		
foot/second	meter/second	$3.048\ 000\ E - 01$
kilometer/hour	meter/second	$2.777\ 778\ E - 01$
knot (International)	meter/second	$5.144\ 444\ E - 01$
mile/hour (U.S. Statute)	kilometer/hour	$1.609\ 344\ E - 04$
VOLUME		
foot^3	meter^3 (m^3)	$2.831\ 685\ E - 02$
gallon (U.S. liquid)	meter^3 (m^3)	$3.785\ 412\ E - 03$
inch^3	meter^3 (m^3)	$1.638\ 706\ 4\ E - 05$
litre	meter^3 (m^3)	$1.000\ 000\ E - 03$
quart (U.S. liquid)	meter^3 (m^3)	$9.463\ 529\ E - 03$
yard^3	meter^3 (m^3)	$7.645\ 549\ E - 01$

GLOSSARY

absorption: A process whereby some of the energy of a seismic wave is converted into heat while passing through a medium (1).

acoustic impedance: Seismic velocity multiplied by density. Reflection coefficient depends on changes in acoustic impedance (1).

A/D: Analog to Digital (1).

air gun: 1. A marine seismic source which injects a bubble of highly compressed air into the water. Oscillations of the bubble as it alternately expands and contracts generate a sonic wave whose frequency depends on the amount of air in the bubble, its pressure, and the water depth (or water pressure). Arrays of guns of different sizes are sometimes used so that a broader frequency spectrum will be generated (1).

alias: An ambiguity in the frequency represented by sampled data. Where there are fewer than two samples per cycle, and input signal at one frequency yields the same sample values as (and hence appears to be) another frequency at the output of the system (1).

amplitude anomaly: An unusually strong amplitude on a seismic section. This may be caused by geometry (i.e., focusing effect at the base of a syncline), constructive (summing) interference, and higher or lower reflection coefficients. A larger reflection coefficient may indicate a zone of faster and higher density (i.e., a hard streak) or a zone of slower velocity and lower density (such as a gas bearing sand) (1).

amplitude modulation (AM): Variations in the amplitude of a high-frequency carrier wave according to low-frequency information (1).

analog: 1. A continuous physical variable (such as voltage or rotation) which bears a direct relationship (usually linear) to another variable (such as the motion of the earth because of seismic waves) so that one is proportional to the other. 2. Continuous, as opposed to discrete or digital (1).

angular unconformity: The lower, older series of beds dip at a different angle to the younger upper beds. This also includes the case where unfolded, younger strata rest upon folded, older strata (1).

anomaly: 1. A deviation from uniformity in physical properties. 2. A portion of a geophysical survey, such as magnetic or gravitational, which is different in appearance from the survey in general. 3. In seismic usage, generally synonymous with structure. 4. Occasionally used for unexplained seismic events (1).

This Glossary has been excerpted from the *Encyclopedic Dictionary of Exploration Geophysics*, by Robert E. Sheriff, published by the Society of Exploration Geophysicists. Permission of the SEG has been received for use of this material.

array processor: A subcomputer controlled by another computer which carries out special functions such as matrix manipulations more efficiently than can be done with the general purpose computer (1).

artificial magnetic anomaly: Local magnetic fields caused by man-made features such as transmission and telegraph lines, electric railways, steel drill casing, pipelines, tanks, etc ... (1).

attenuation: 1. A reduction in amplitude or energy, such as might be produced by passage through a filter. 2. A reduction in the amplitude of seismic waves, such as produced by divergence, reflection and scattering, and absorption. 3. That portion of the decrease in seismic or sonar signal strength with distance not dependent on geometrical spreading. This decrease depends on the physical characteristics of the transmitting media, involving reflections, scattering and absorption (1).

automatic gain control (AGC): A system in which the output amplitude is used for automatic control of the gain of an amplifier. Seismic amplifiers usually have individual AGC for each channel, although multichannel control is sometimes used (1).

average velocity: The distance traversed by a seismic wavelet divided by the time required, both often corrected to a reference datum plane. For reflections, often refers to a ray reflected at normal incidence (1).

background: Average noise level, whether systematic or random, upon which a desired signal (such as a reflection) is superimposed. Usually refers to the total noise, independent of the presence of the signal (1).

band pass: The range of frequencies which a system passes with little attenuation, while substantially attenuating frequencies outside these limits (1).

band reject filter: A filter which attenuates a range of frequencies. May be considered the inverse of a band pass filter (1).

bar: 10^5 N/m^2 (newtons per square meter). A unit of pressure, approximately one atmosphere or 14.5 psi (1).

base line: The line between two radio positioning base stations which form a pair (1).

basement: 1. Geologic basement is the surface beneath which sedimentary rocks are not found; the igneous, metamorphic, or granitized, or highly folded rock underlying sedimentary rocks (1).

BCD: Binary coded decimal (1).

beam width: The beam angle of an acoustic transducer, typically measured between the 3dB points on a graph of the intensity vs. angle. By convention, in side scan sonar, the beam width is the combined beam width of transmission and reception (2).

bedding plane: The division planes which separate individual sedimentary layers, beds, or strata (3).

binary: Composed of only two elements. A number system in which two digits, 0 and 1, are used, the position of the digits representing the powers of two (1).

bird: A hydrophone cable depth controller (1).

Boomer: A magnetostrictive type seismic source employed in HRG surveys. EG&G Trademark (4).

bright spot: An event on a seismic section attributable to gas accumulation. Ideally characterized by an anomalously strong amplitude, phase changes at either end, attenuation of reflections, and lower frequencies immediately below, etc.

bubble pulses: 1. Pulses set up by oscillation of a bubble of high-pressure gas in the water. 2. Repetitions on a seismic record of the first arrivals and all other shot-generated events because of oscillations of the bubble (1).

bust: Failure to tie a survey loop within acceptable standards (1).

CSP (Continuous Seismic Profile): Usually refers to single channel analog recording.

cable: The assembly of electrical conductors used to connect the hydrophones to the survey vessel. Also see **streamer** . . . (1).

cable strum: Vibrations of a marine streamer (or tow fish cable) produced by sudden tension such as might be caused by pitching of the towing ship or jerks from the tail buoy. A source of noise in marine seismic data . . . (1).

casing: Steel pipe or tubing installed about a drill hole to prevent it from caving in, and isolating the drill string from the environment (i.e., water column, between drilling rig and sea floor).

casing point: The depth to which initial casing is set and cemented during the drilling of the well. The depth may be planned or be necessitated by subsurface conditions such as the presence of shallow gas or faulting.

cavitation: The situation where the pressure in a liquid becomes lower than the hydrostatic pressure, often reaching the vapor pressure of the liquid. The collapse of liquid into the region produces very large pressures and generates a shock wave by implosion (1). In the acquisition of HRG seismic data, cavitation beneath a transducer results in a loss of recorded signal, due to the rolling motion of the vessel in rough seas.

channel cut and fill: Relict (buried) river or stream channels filled with characteristically dipping strata.

CHIRP: A Vibroseis signal; a sinusoidal signal of continuously varying frequency. Often implies a linear change of frequency with time (1).

collapse depression: Semicircular area of the sea floor, up to several hundred meters in diameter, produced as a result of mass sediment displacement. Generally a precursor to the formation of a mud flow gully.

common-depth-point (CDP): The situation where the same portion of the subsurface is involved in producing reflections from a common mid point between source and receiver locations. The summation of these points, after correcting for normal moveout, produces a common-depth-point stack (1).

compressional wave: An elastic body wave in which particle motion is in the direction of propagation; the type of seismic wave assumed in conventional (and HRG) seismic exploration. Also called: P Wave, dilatational wave, and longitudinal wave (1).

conformable: Two adjacent parallel beds separated by a surface of original deposition, where no disturbance or denudation occurred during their deposition (1).

consolidation: 1. In soil mechanics: The adjustment of a saturated soil in response to increased load (overburden). Involves the squeezing of water from the pores and decrease in void ratio. 2. In geology any or all of the processes whereby loose, soft, or liquid earth materials become firm and coherent (3).

contour: A line connecting points of equal value, or representing the locus of a constant value on a map or diagram (1).

contour interval: The difference in value between two adjacent contour lines. Abbreviated CI (1).

corer: A device for taking sediment or rock samples from the sea floor. Numerous types exist of which the following are commonly used in association with HRG surveys: box corer, gravity corer, piston corer, vibra-corer, and grab sampler.

correlation: 1. Identifying a phase of a seismic record as representing the same phase on another record indicating that events on two seismic records are reflections from the same stratigraphic sequence, or refractions from the same marker. 2. The degree of linear relationship between a pair of traces; a measure of how much two traces look alike or the extent to which one can be considered a linear function of the other (1).

crossfeed: Interference resulting from the unintentional pick up by one channel of information or noise on another channel (1). Phenomena in which two sides of an acoustic system, such as a side scan sonar, interfere with each other (2).

crosstalk: See **crossfeed.**

cutoff: The frequency at which a filter response is down by a pre-determined amount, such as 3 dB. The cuttoff points designate the filter; e.g., an 18-57 filter has a low-frequency cutoff at 18 Hz and a high-frequency cutoff at 57 Hz (1).

D/A: Digital to analog (1).

datum: 1. The arbitrary reference level to which measurements are corrected. 2. The reference level for elevation measurements, often sea level (1).

dB: Decibel (1).

deconvolution: The process of undoing the effect of another filter. Usually an inverse filter is designed and convolved with the signal, the objective being to nullify an objectional effect of an earlier filter action (1).

delta t: moveout, usually written Δt. The time difference between the arrival times at different geophone (hydrophone) groups (1).

demodulation: The process of retreiving an original signal from a modulated signal (1).

demultiplex: Separating the individual component channels which have been multiplexed (1).

depth controller: A device with moveable vanes which fastens to a marine streamer to maintain it at a predetermined depth (1).

diapir: A flow structure whose mobile core has pierced overlying rocks. Salt and shale are the most common sedimentary rocks involved in diapirs. Intrusive rocks can also form diapirlike features but diapirism is usually restricted to plastic flow (1).

diffraction: 1. A phenomenon common to all waves (light, radio, seismic, water surface, etc.) 2. Scattered seismic energy which emanates from an abrupt discontinuity of rock type, particularly common where faults cut reflecting interfaces. Diffractions from such discontinuities below the seismic line appear along a diffraction curve (which depends upon the velocity distribution above the diffracting point) (1).

digital: Representation of quantities in discrete (quantized) units (1).

dip: 1. The angle which a plane surface makes with the horizontal. 2. The angle which bedding makes with the horizontal. 3. The angle which a reflector or refractor makes with the horizontal (1).

diurnal variation: Daily fluctuations of the geomagnetic field, related principally to the tidal motion of the ionosphere. Amplitude and phase vary with season and latitude by as much as 40 γ (1).

dog house: The hut (or cab) which contains seismic recording instruments in the field (aboard ship) (1).

doppler sonar: A sonic location system used by ships, based upon the Doppler effect . . . The ship's velocity is computed from the frequency shift of the reflected (sound) beams (1).

dynamic range: 1. The ratio of the largest recoverable signal (for a given distortion level) to smallest recoverable signal. The smallest recoverable signal is often taken to be the noise level of the system, and dynamic range is often so defined. However, signals often can be extracted even though they are buried in the noise. The ambiguity can be solved if the entire signal extraction process is considered in defining dynamic range rather than considering the recording equipment only. 2. The ratio of the maximum reading to the minimum reading (often noise level) which can be recorded by and read from an instrument without change of scale (1).

echo sounder: A device for measuring water depth by timing sonic reflections from the water bottom; a fathometer (1).

electromechanical source: A subbottom profiler. A marine seismic energy source in which capacitors are charged to high voltage and then discharged through a transducer in the water.

electrostatic graphic recorder: An instrument which displays analog data (such as a sparker or subbottom profiler) on heat sensitive paper. Dark areas are burned into the paper; the darkness varies with the incoming signal intensity and various instrument settings (6).

engineering geophysics: Use of geophysical methods to get information for civil engineering. Bathymetry, seismic reflection and refraction, side scan sonar, gravity, magnetic, electrical, and sampling methods are employed. In water covered areas highpower fathometers, sparkers, gas guns, and other seismic reflection methods employing high frequencies (up to 5 kHz) are used to obtain reflections from shallow interfaces so that bedrock and the nature of fill material can be diagnosed. Such methods are also used to locate large pipelines on, or buried in, the sea bottom by the prominent diffractions which they generate (1).

equalization: 1. Trace equalization involves adjusting the gain of different channels so that the amplitudes are comparable (1).

event: A lineup on a number of traces which indicates the arrival of new seismic energy, denoted by a systematic phase or amplitude change on a seismic record; an arrival. May be a reflection, refraction, diffraction or any other type of wavefront (1).

fathometer: A device for measuring water depth by timing sonic reflections from the water bottom; an echo sounder (1).

fault: A displacement of rocks along a shear surface. The surface along which the displacement occurs is called the fault plane (often a curved surface and not a plane in the geometric sense). The dip of the fault plane is the angle which it makes with the horizontal; the angle with the vertical is called the hade. The trace of a fault is the line which the fault plane makes with a surface (often the sea floor). Faults are classified as normal, reverse, or strike slip, depending on the relative motion along the fault plane. Faulting during sediment deposition may cause stratigraphic associations with faulting. Thus beds may abruptly thicken and become more sandy across a normal **growth fault**. Faults evidence themselves in seismic data principally by: a. abrupt termination of events, b. diffractions, changes in dip, flattening or steepening, d. distortions of dips seen through the fault, e. cut out of data beneath the fault or in the shadow-zone of the fault, f. changes in the pattern of events across the fault, and g. occasionally a reflection from the fault plane (1).

ferromagnetic: Having positive and relatively large magnetic susceptibility and generally large hysteresis and remanence . . . (1).

field tape: A magnetic tape containing geophysical data recorded in the field, as opposed to a processed tape on which the data have been modified by processing in a computer (1).

filter: 1. That part of a system which discriminates against some of the information entering it. The discrimination is usually on the basis of frequency, although other bases such as wavelength or moveout, or coherence may be used . . . (1).

firing rate: Timing interval between successive shots or firing of a seismic source, usually controlled by the recording system for analog systems (6).

first break: The first recorded signal attributable to seismic wave travel from a known source. First breaks on reflection records are used for information about the weathering (land work). Much refraction work is based principally on first breaks, although secondary (later) refraction arrivals are also used. Also called first arrival (1).

fish: A sensor which is towed in the water, such as used with side scan sonar (1).

fix: A determination of location, as by the intersection of two lines-of-position (1).

fixed point: A method of calculating in which the computer does not consider the location of the decimal point. Data must have values within a certain range which is the responsibility of the programmer. Compare **floating point** (1).

floating point: A number expressed by a certain number of significant figures times a base raised to a power . . . (1).

fluxgate mangetometer: An instrument capable of detecting changes in the geomagnetic field on the order of one gamma . . . (1).

FM: Frequency modulation (1).

fold: Common-depth-point multiplicity, where the same CDP point is sampled at 12 offsets, e.g. it is referred to as twelve-fold (1).

frequency: Symbol, f. The repetition rate of a periodic waveform, measured in per second or hertz. The reciprocal of period . . . (1).

gain: An increase (or change) in signal power (or amplitude) from one point in a circuit or system to another (1).

galvanometer: A device to measure small currents. A coil is suspended in a constant magnetic field which rotates through an angle proportional to the electric current flowing through the coil. A part of many seismic cameras. Often abbreviated galvo (1).

gamma: A unit of magnetic field. A gamma $= 10^{-5}$, gauss $= 10^{-9}$, telsa $= 1$ nanotelsa (1).

gas hydrate: a. Hydrated methane (mainly methane with occasional small amounts of other hydrocarbons) in a solid state. b. A crystalline icelike clathrate compound, which under certain conditions of low temperature and high pressure become solid, occasionally forming seismic reflectors (7).

gas seep: Active entry of gas bubbles into the water column. The gas may be of shallow-to-deep origin (6); and may be detected on HRG subbottom and echo sounder records.

gate: The interval of record time over which a function (such as autocorrelation or crosscorrelation) is evaluated. Also called window (1).

gather: A display of the input data to a stacking process rearranged so that all the seismic traces corresponding to some criterion are displayed side-by-side. Used for checking corrections and evaluating the components of the stack . . . (1).

geophysical survey: A program to make geophysical measurements over an area with the objective of learning about the geology (1). The objective of many **HRG surveys** is to locate hazards or constraints to the conduct of exploration and production operations, as well as locate mineral deposits such as sands and gravels.

geophysics: 1. The study of the earth by quantitative physical methods, especially by seismic reflection and refraction, magnetic, electrical, and radiation methods. 2. The application of physical principles to studies of the earth (1).

geostatic pressure: Pressure due to the weight of the overlying rock, which will generally be different from the pressure of fluids in the rock (1).

ghost: 1. Energy which travels upwards from the shot and then is reflected downward, such as occurs at the base of the weathering or at the (sea) surface . . . (1).

gradiometer: 1. A device for measuring the gradient of a potential field. 2. An arrangement of two magnetometers, one above the other, so that the difference in their readings is proportional to the vertical gradient of the magnetic field . . . (1).

ground: A point in a cricuit used as a common reference point, often the conducting chassis on which the electrical circuit is physically mounted. Frequently, but not necessarily, connected to the earth by a low-resistance conductor (i.e., the vessel's hull) (1).

group: The various geophones (or hydrophones) which collectively feed a single channel . . . (1).

group interval: The horizontal distance between the centers of adjacent geophone (or hydrophone) groups (1).

hardware: Equipment, especially computing machine equipment (1).

harmonic: A frequency which is a simple multiple of a fundamental frequency. The third harmonic, for example, has a frequency three times that of the fundamental . . . (1).

head: A device which reads from or writes on a storage medium (1), (i.e., magnetic head for magnetic tape).

Hertz: A unit of frequency. Hertz = Hz = cycles per second = cps (1).

high-cut filter: A filter that transmits frequencies below a given cutoff frequency and substantially attenuates all others. The same as a low-pass filter (1).

high-pass filter: A filter that passes without significant attenuation frequencies above some cutoff frequency while attenuating lower frequencies. The same as low-cut filter (1).

high-resolution geophysics (HRG): That field of geophysics concerned with obtaining fine details, especially near-surface details; usually implies the recording of higher frequencies (greater than 100 Hz) than normal.

horizon: The surface separating two different rock layers, where such a surface (even though not itself identified) is associated with a reflection which can be carried (traced) over a large area; a map based upon the reflection event may be called a horizon map . . . (1).

horst: A block raised up by normal faulting; usually long compared to its width (1).

HR: High resolution; recording of seismic frequencies (up to 500 Hz, but especially 75 to 150 Hz) above the normal exploration range with the objective of improving resolution, especially of shallow events (1).

hydrophone: A pressure detector, a detector which is sensitive to variations in pressure, as opposed to a geophone which is sensitive to motion. Used when the detector can be placed below a few feet of water as in marine or marsh work . . . (1).

implosion: Collapse into a region of very low pressure. The creation of such a region underwater (as with Flexichoc, trademark, I.F.P.) causes the water to rush in with great force and the collision of the inrushing water on itself generates a seismic shock wave. The outrush of water propelled by a bubble of high-pressure gas (as from an underwater explosion or from an air gun) leaves behind a region at the vapor pressure of the water, into which the water subsequently collapses, causing the bubble effect (1).

incident angle: The angle which a raypath makes with a perpendicular to an interface, which is the same as that which a wavefront makes with the interface in isotropic media (1).

in situ: Material in its original location . . . (1).

integrated navigation system (INS): A combination of positioning systems in a synergetic manner. Specifically, the combination of satellite navigation with Doppler sonar and gyrocompass (and other subsystems) into a system . . . (1).

interpreter: 1. One who determines the geological significance of geophysical data. 2. A machine which reads coded information (such as punched cards) and prints out the translation. 3. A computing machine routine which translates from one computer language into machine language at execution time, as programs written in Fortran (1).

interstitial water: Water in the interstices or pore spaces of a formation (or sediment) (1).

interval velocity: Seismic wave velocity measured over a depth interval . . . (1).

isopach: 1. A contour which denotes points of equal thickness of a rock type, a formation, a group of formations, etc. Usually used for isochore. 2. A contour denoting equal vertical distance, thus not necessarily corrected for the dip of the bedding. 3. Sometimes incorrectly used for a contour denoting equal time difference between reflections. Compare **isotime** (1).

isotime: 1. The time interval between two reflections, mapped in order to locate stratigraphic changes in the interval, reef buildup, variation in salt thickness, etc. The isotime may vary because the velocity or the thickness varies or both. 2. Contours of reflection time, time intervals, etc. . . . (1).

isotropic: Having the same physical properties regardless of the direction in which they are measured . . . (1).

JOIDES: Joint Oceanographic Institutions for Deep Earth Sampling. A program to obtain cores of the sediments in the deep oceans. Holes drilled from the ship *Glomar Challenger* did much to prove plate tectonics and hence has had a tremendous impact on geology and geophysics (1).

Julian day: The number of a day within a calendar year. Referred to Greenwich (1).

k: 1. Wave number; the number of waves per unit distance; the reciprocal of wavelength (1).

key bed: A reflection with sufficient distinguishing characteristics to make it easily identifiable in correlations (1).

kick: 1. Break or onset. 2. Loss of drilling fluid because fluid is penetrating a porous formation. 3. Gas entering the drilling fluid from a porous formation (1).

lane: The unit of measuring position with radio positioning systems (RPS). In phase comparison (CW) systems, the lane is the distance represented by one cycle of the standing wave interference pattern resulting from two radiated waves. Its linear distance is not constant but depends on the position within the network. Phase comparison systems permit location within a lane but do not necessarily determine in which lane . . . (1).

lateral variations: Changes in a horizontal direction (1).

leakage: Low electrical resistance to ground where there should be high resistance, as in a wet seismic cable (or sensor tow fish cable) (1).

line: 1. A linear array of observation points, such as a seismic line . . . (1). 2. Seamanlike term for a landlubber's rope.

Lloyd mirror effect: Light and dark bands extending along the sonograph, whose separation increases with distance from the ship and with increase in water depth. The pattern is only produced in shallow water during calm weather. It is caused by the interference between sound following a direct path from the sea floor to the transducer, with that following a longer route involving a reflection at the sea surface (5).

lobes: Passbands in a directivity graph. A bulge in a spectral window. The main window is called the main lobe and smaller bulges to the side are called side lobes. Used in connection with seismic directivity, transducer patterns, etc.... (1).

log: 1. A record of measurements or observations, especially those made in a borehole. 2. An instrument for measuring a vessel's speed or distance traveled or both (1).

low-cut filter: A filter that transmits frequencies above a given cutoff frequency and substantially attenuates lower frequencies. Same as high-pass filter (1).

low-pass filter: A filter that passes frequencies below some cutoff frequency while substantially attenuating higher frequencies. Same as high-cut filter (1).

MES: Multiple electrode sparker.

magnetic resonance: Interaction between the magnetic moments (electron spin and/or nuclear spin) of atoms with an external magnetic field. Magnetic resonance is basic to the operation of the proton magnetometer and optically pumped magnetometer (1).

magnetic signature: The shape of a magnetic anomaly (1).

magnetic storms: Rapid, irregular, transient fluctuations of the magnetic field which are greater in magnitude, more irregular, and more rapid than diurnal variations. These occur most commonly during unusual sunspot activity as a result of bombardment of the earth by high-energy particles from the sun. Magnetic storms commonly have amplitudes of 50 to 100 γ, occasionally thousands of gammas. Magnetic prospecting usually has to be suspended during such periods (1).

magnetometer: An instrument for measuring magnetic field strength ... (1).

magnetostriction: Change in the strain of a magnetic material as a result of changes in magnetization ... (1).

main lobe: 1. The portion of a directivity graph which indicates the continuous band of directions (or apparent wavelengths) in which the greatest energy is radiated or which undergo least attenuation ... (1). 2.

The portion of an acoustic beam containing most of the energy emitted by a transducer (5).

map: 1. To transform information from one form to another or, 2. the product of such transformation . . . (1).

Mercator projection: A conformal cylindrical map projection developed on a cylinder tangent along the equator with the expansion of the meridians equal to that of the parallels (1).

microsecond: 10^{-6} seconds.

migration: Plotting of dipping reflections in their true spatial positions rather than directly beneath the point midway between the shotpoint and center of the geophone (hydrophone) spread . . . (1).

millisecond: A thousandth of a second; abbreviated msec (1).

mistie: 1. The difference obtained on carrying a phantom or reflection or some other measured quantity around a loop. 2. The difference of values at identical points on intersecting lines or of values determined by independent methods . . . (1). Also, misclosure.

monitor record: A record made as a check. A record made in parallel with a magnetic tape recording at a time of a shot. 2. A record of shipboard gravity and magnetic measurements made while the measurements are being obtained (1).

monument: 1. An identifiable point on the ground to which surveys can be tied. May be an inscribed tablet on concrete, a steel fence picket with identification attached, etc. Also: bench mark (1).

moveout: Stepout. The difference in arrival time at different geophone (hydrophone) positions. Arrival times differ because of, a. normal moveout, differences because of variable shotpoint-to-geophone (hydrophone) distance along the reflection paths, b. dip moveout, differences because of reflector dip, and c. statics, differences because of elevation (depth) and weathering variations. See also delta t . . . (1).

mud (drilling): An aqueous suspension used in rotary drilling. Mud is pumped down through the drill pipe and up through the annular space between it and the walls of the hole. The most common bases of drilling muds are bentonite, lime, and barite in a finely divided state. The mud helps remove cuttings, seal off porous zones, and hold back formation fluids (1).

mud line: Sea floor; usually used in conjunction with engineering cross sections, and borehole data.

mud lumps: Diapirs of mud which rise up through sediments in areas of rapid sedimentation at the mouth of major distributaries, i.e., South Pass

Area of Mississippi River Delta. Mud lumps are probably an intrusion of older shelf and prodelta clays into and through overlying bar deposits (8).

mud volcano: . . . A more or less violent eruption or surface extrusion of watery mud or clay which almost invariably is accompanied by methane gas, and which commonly tends to build up a solid mud or clay deposit around its orifice which may have a conical or volcanolike shape (9).

multiple: Seismic energy which has been reflected more than once . . . (1).

multiple coverage: Seismic arrangement whereby the same portion of the subsurface is involved in several records, as with CDP shooting. The redundancy of measurements permits various types of noise to be attenuated in processing (1).

multiplex: A process which permits transmitting several channels of information over a single channel without crossfeed . . . (1).

mute: To change the relative contribution of the components of a (seismic) record stack with record time. In the early part of the record the long offset (far) traces may be muted or excluded from the stack because they are dominated by refraction (also first arrivals) or because their frequency content after NMO correction is appreciably lower than other traces . . . (1).

near-trace gather: A record section which comprises only the data from the geophone (hydrophone) group (or few groups) nearest the shot (source) (1).

neutrally buoyant: Having the same buoyancy as the fluid in which it is immersed. Seismic streamers are usually very nearly neutrally buoyant so that very little force is required to submerge or raise them (1).

normal moveout (NMO): The variation of reflection arrival time because of variaton in the shotpoint-to-geophone (hydrophone) distance (offset) . . . (1).

notch filter: A filter which is designed to remove a single narrow band of frequencies. Often used to remove high line effects (60/50 Hz) (1).

Nyquist frequency: A freqeuncy associated with sampling which is equal to half the sampling frequency . . . (1).

observer: The person in charge of the recording on a seismic crew. Sometimes the observer is also the field manager; sometimes he/she is principally an electronic technician . . . (1).

Occam's Razor: The simplest explanation of observations is the most probable. A dictum of scientific reasoning (1).

octave: Separation of two frequencies having a ratio of 2 (or ½) . . . (1).

offset: The distance from the shotpoint to the center of the nearest geophone (hydrophone) group . . . (1).

one-way time: Half the corrected traveltime for a reflection arrival. One-way time multiplied by average velocity gives reflector depth for a flat reflection and flat velocity layering (1).

open hole: A wellbore that has not been cased, where measurements are made (1).

overburden: 1. Material lying over an ore or valuable deposit. 2. The section above a refractor or above a reflector (1). 3. Weight, usually buoyant, of soil or sediment column above a particular sample depth.

over consolidated: See consolidation.

party: The group of persons working together to carry out a geophysical field project. Also called crew (1).

party chief: The person in charge of a geophysical party (1).

party manager: The person working under the party chief who usually is responsible for the field work (1).

peak: The maximum upward (positive) excursion of a seismic wavelet. Opposite of trough (1).

peg-leg multiple: 1. A multiple reflection involving different interfaces so that its travel path is nonsymmetric . . . 2. A short path multiple which is repeatedly reflected within thin formations . . . (1).

penetration: The greatest depth from which seismic reflections can be picked with reasonable certainty . . . (1).

permeability: 1. The ratio of the magnetic induction (B) to the magnetic intensity (H) . . . 2. A measure of the ease with which a fluid can pass through the pore spaces of a formation. Measured in millidarcy units . . . (1).

phantom: The line on a seismic section drawn parallel to the dip of nearby reflection events. Phantoms are drawn and mapped where one cannot follow one event far enough to develop a map on that event alone (1).

phase: The angle of lag or lead (or displacement) of a sine wave with respect to a reference . . . (1).

phase inversion: A change of 180° in phase angle . . . (1).

pick: To select an event on a seismic record, as to pick reflection events . . . (1).

piezo-electric: The property of a dielectric which generates a voltage across it in response to stress, and vice-versa. In a hydrophone the stress is produced by the pressure. Piezo-electric transducers are commonly made of barium titanate or zirconate (1).

pinger: 1. A transponder or device which emits an acoustic signal upon being activated by sensing a coded acoustic signal. Pingers placed on the sea bottom or in anchored buoys can be interrogated by a ship transmitting the coded acoustic (sonar) signal and the distance to the pinger determined by the traveltime measurements. 2. A shallow penetration high-power transducer used in marine engineering studies in soft bottom areas, also subbottom profiler (1).

pockmarks: Cone-shaped depressions, sometimes 5 to 10 m deep and 15 to 45 m in diameter. "They were possibly formed by ascending gas or subsurface water leakage from underlying coastal plain sediments (10)."

porosity: Pore volume per unit volume of formation . . . (1).

positioning: Determining the location of a survey ship or aircraft, usually with respect to geodetic coordinates but sometimes with respect to reference beacons which may not be at known locations . . . (1).

postplot: Computation of positioning locations which have been previously occupied, based on the best reconciling of all available data (1).

pot hole: A broad, shallow depression in the sea floor, typically it is roughly circular, 50 ft to 200 ft in diameter, and 2 ft to 15 ft deep. It probably results from sediments collapsing after gas has gently escaped (6). See also **collapse depression**.

predictive deconvolution: Use of information from the earlier part of a seismic trace to predict and deconvolve the latter part of the trace . . . (1).

preplot: 1. The location programmed for shotpoints (or boat position) before the points are actually occupied. 2. A list of programmed points in radio navigation system coordinates . . . (1).

primary reflection: Energy which has been reflected only once and hence is not a multiple . . . (1).

processing: Changing the form of data. The objective of processing is usually to improve the signal-to-noise ratio so as to facilitate interpretation of the data . . . (1).

profile line: The line along which measurements are made (1).

profiler: A marine seismic reflection system usually involving a low energy source such as a sparker and only one or two hydrophone groups . . . (1).

pulse: A waveform whose duration is short compared to the time scale of interest and whose initial and final values are the same (usually zero) . . . (1).

P wave: An elastic body wave in which particle motion is in the direction of propagation. The type of wave assumed in conventional seismic exploration (and HRG). Also called compressional wave, longitudinal wave, primary wave, pressure wave, dilatational wave, and irrotational wave . . . (1).

QC: Quality control (1).

quenching: White lines across the sonograph, which result from air bubbles in the water blanketing sound transmission and reception from the transducer. This effect arises when a ship is pitching and rolling in bad weather (5).

ramp: To change from one set of parameters to another, usually implying in a linear manner . . . (1); i.e., the change in gain from an onset value of zero to some higher gain level with respect to time.

raypath: A line everywhere perpendicular to wavefronts (in isotropic media) . . . (1).

real time: Having the same time scale as actual time. Processing of seismic data at the same rate as that at which they were recorded . . . (1).

record: A recording of the energy from one shot (or other type of energy release) picked up by a spread of geophones (hydrophones). May be on photographic or other paper or on magnetic tape . . . (1).

record section: Display of seismic traces side-by-side to show the continuity of events . . . (1).

reflection: The energy or wave from a shot or other seismic source which has been reflected (returned) from an acoustic impedance contrast (reflector) or series of contrasts within the earth . . . (1).

reflection coefficient: The ratio of the amplitude of the displacement of a reflected wave to that of the incident wave . . . (1).

reflection survey: A program to map geologic structure employing the seismic reflection method . . . (1).

reflector: A contrast in acoustic impedance, which gives rise to a seismic reflection . . . (1).

refraction: The change in direction of a seismic ray upon passing into a medium with a different velocity . . . (1).

relict: Geologic features which have been formed in the past and which are preserved in places where they cannot form at present; i.e., buried river channels at and below the present sea floor (3).

resolution: The ability to separate two features which are very close together . . . (1).

reverberation: Multiple reflection in a layer, usually the water layer in marine work . . . (1).

ringing: Reverberation or singing; the oscillatory effect on seismic waveforms produced by short path multiples in a shallow water layer . . . (1).

sample interval: The interval between readings, such as the time between successive samples of a digital seismic trace. Also called sample period (1).

SatNav: Satellite navigation (1).

sea chest: A fitting in a ship's hull below the waterline, such as used to mount sonar transducers (1).

sea state: A standardized description of marine weather and wave conditions, which are useful toward analyzing sources of noise on HRG seismic records. Several different systems are currently in use (i.e., Beaufort Scale).

section: What might be seen by slicing through a solid object such as a slice through the earth . . . A plot of seismic events, as in a record section . . . (1).

secular variation: Time variations whose periods are measured in decades, as "the secular variation of the earth's magnetic field" with periods of thirty to three hundred years (1).

seismic: Having to do with elastic waves . . . (1).

seismic survey: A program for mapping geologic structure by creating seismic waves and observing the arrival time of the waves reflected from acoustic impedance contrasts or refracted through high-velocity members . . . (1).

seismogram: A seismic record (1).

shallow water survey: A geophysical survey in waters where conventional marine survey ships cannot operate easily because of shallow water, reefs, or other obstructions, etc. (1).

shear strength: The shear strength of a sediment is a function of its cohesion plus the product of the effective load, normal to the shear plane and the tangent of the effective internal friction angle.

shear wave: A body wave in which the particle motion is perpendicular to the direction of propagation. Also called S-wave or transverse wave ... (1).

shotpoint: Abbreviated SP. The location where an explosive charge, or other source of seismic energy, is detonated or fired ... (1).

side lobes: Secondary (relative to the main) acoustic transmission beams.

side scan sonar: A method of locating irregularities on the ocean bottom. A pulse of sonar energy (typically 120 kHz) is emitted from a fish which is towed 50 to 500 ft above the bottom, depending on the range and resolution sought. The sonar beam is narrow in the direction of traverse because the source consists of a line array of elements. Bottom irregularities (rock outcrops, pipelines, shipwrecks, boulders, and variations in bottom sediments produce changes in the amount of energy return. The arrival time measures the distance from the fish to the reflecting object (1).

Other names include: Asdic, basdic, sideways asdic, sideways looking sonar, sideways sonar, echo ranger, horizontal echo sounder, and lateral echo sounder (5).

signal-to-noise ratio: The energy of desired events divided by all remaining energy (noise) at that time. Often abbreviated S/N ... (1).

signature: Waveshape which is characteristic of a particular source, transmission path, or reflecting sequence (1).

skew: A condition that occurs when a magnetic tape is not mounted properly with respect to the magnetic heads. The tracks may deviate from their proper position and produce crossfeed, time displacement between channels, parity errors, etc. (1).

sky wave interference: Interference between the direct (or ground) radio wave and waves reflected from the ionosphere ... Sky wave interference degrades the accuracy of radio positioning systems (1).

sleeve exploder: A seismic marine energy source in which a gas (propane or butane) is exploded in a thick rubber bag (the sleeve), and from which the waste gasses are vented directly to the air rather than into the water, thus reducing the bubble effect. Also called Aquapulse. Esso Production Research Company patent (1).

slump: A rotational slide; a downward and outward movement of a portion of the soil or soil rock mass ... (11).

Snell's law: When a wave crosses a boundary between two isotropic media, the wave changes direction such that the sine of the angle of incidence (angle between the wavefront and a tangent to the boundary) divided by the velocity in the first medium equals the sine of the angle of refraction divided by the velocity in the second medium . . . (1).

sonic log: A well log of the traveltime (transit time) for acoustic waves over a unit distance, and hence the reciprocal of the longitudinal wave (P-wave) velocity . . . (1).

sonobuoy: A device used in marine refraction surveys for detecting energy from a distant shot and radioing the information to the recording ship . . . (1).

sonograph: An acoustic picture obtained underwater by means of oblique illumination with side scan or sector scanning sonars (5).

Sonoprobe: A type of marine echo sounder that generates sound waves and records their reflections. Usually has more power and more penetration than a fathometer but less than a sparker or gas exploder. Mobil Oil Co. trademark (1).

sparker: A seismic source in which an electrical discharge in water is the energy source (1).

speed: Used synonomously with velocity in seismic work (1).

spherical divergence: The decrease in wave strength (energy per unit of area of wavefront) with distance as a result of geometric spreading . . . (1).

squiggle: A wiggle trace or trace of a glavanometer deflection versus time . . . (1).

stack: A composite record made by mixing traces from different records . . . (1).

statics: Corrections applied to seismic data to eliminate the effects of variations in elevation, weathering thickness, or weathering velocity. The objective is to determine the reflection arrival times which would have been observed if all measurements had been made on a (usually) flat plane with no weathering or low-velocity material present . . . (1).

stepback: The correction applied to a location (such as the location of a seismic ship determined by radio methods) to yield the mid point of the subsurface coverage for a seismic record, allowing for the position of the cable and the shot with respect to the navigation antenna (1). Also called: set back or lay back.

strain: The change of shape associated with stress of a body (1).

streamer: A marine cable incorporating pressure hydrophones internally, designed for continuous towing through the water. A marine streamer is typically made up of 24 active or live sections which contain hydrophone arrays separated by spacer or dead sections. Usually the streamer is nearly neutrally buoyant and depressors or depth controllers are attached to depress the streamer to the proper towing depth. The entire streamer may be up to 5000 to 9000 ft in length (1).

stress: The intensity of force acting on a body, in terms of force per unit area (1).

subbottom profiler (SBP): Any of various acoustic instruments which produce a cross-sectional-like record of sediments from the sea floor down to as much as several hundred feet. Abbreviated: SBP.

survey: To determine the form, extent, position, subsurface characteristics, etc., of an area or prospect by topographical, geological, or geophysical measurements (1).

sweep rate: The speed at which data are recorded on a graphic recorder; it determines the time scale of the presentation (i.e., full scale being 0.25 sec, 0.5 sec, 1.0 sec, etc.) (6).

tap test: A recording made as a geophone (hydrophone) is tapped lightly, showing which channel the geophone feeds. Used to check that the spread (streamer) is properly connected and oriented and also that the geophone is live (1).

thermocline: The decrease in water temperature with depth in the ocean . . . (1).

throw: The vertical component of separation of a bed by a fault (1).

tie line: A survey line which connects other survey lines (usually at an angle of 90°) . . . (1).

tilt angle: The angle between the mid line of the main beam and the horizontal (for an acoustic transducer) (5).

time break: The mark on a seismic record which indicates the shot instant or the time at which the seismic wave was generated (1).

time variable gain (TVG): A signal processing technique used in underwater acoustics to adjust for attenuation in order to try to produce uniform results. The curve of gain vs. time varies with the type of system and the signal situation. Some systems use bottom triggered TVG in order to produce a clean water column. This can also eliminate echoes from gas seeps or fish.

timing lines: Marks or lines at precise intervals of time such as used on HRG seismic records . . . to help measure the arrival times of seismic events . . . (1).

towing depressor: A device used to pull down, to greater depths than when strung out, a side scan sonar fish or other towed vehicle.

trace: 1. A record of one seismic channel, electromagnetic channel, etc . . . 2. A line on one plane representing the intersection of another plane with the first one, such as a fault trace (1).

trace equalization: Adjusting a seismic channel so that the amplitudes of adjacent traces are comparable in the sense of having the same rms value over some specified interval or some other criterion (1).

transceiver: Device which is both a transmitter and receiver, such as used in sonar (1).

transducer: 1. A device which converts one form of energy into another. Many types of transducers are reverisble, i.e., converting electrical energy into acoustical energy and vice-versa . . . (1).

transponder: A device which transmits a signal upon receiving another signal, used with both electromagnetic waves and sonar acoustic waves. Also called a pinger (1).

unconformity: A surface of erosion or nondeposition that separates younger strata from older rocks. An unconformity ordinarily is not recognizable in seismic data except where the layers above and below the unconformity are not parallel (angular unconformity) (1).

Uniboom: A marine seismic energy source in which capacitors are charged to high voltage and then discharged through a transducer in the water. The transducer consists of a flat coil with a spring loaded metal plate. Eddy currents force the plate to separate sharply from the coil, generating an acoustic pressure wave. EG & G trademark.

variable area: A display method in which the width of a blacked-in area is roughly proportional to the signal strength (1). Similar to variable density display.

velocimeter: A instrument for measuring the velocity of sound in water, used to correct Doppler sonar data for salinity and temperature variations . . . (1)

velocity: 1. A vector quantity which indicates time rate of change of displacement. 2. Usually refers to the propagation rate of a seismic wave without implying any direction. Velocity is a property of the medium and is not a vector quantity when used in this sense; speed . . . (1).

velocity analysis: Calculation of velocity from measurements of normal moveout . . . (1).

vented gas-charged sediment cone: A full column of gas bubbles above a sea-floor mound which is caused by gas from depth, forcing its way to the sea bottom.

vertical exaggeration: 1. The use of a vertical scale which is larger than the horizontal scale. Exaggeration makes subtle effects more evident but distorts structural relationships. Seismic time sections involve variable vertical exaggeration because the velocity varies with depth. The picking and interpretation of significant features on record sections is greatly affected by vertical exaggeration . . . 2. The ratio of vertical scale to horizontal scale (1).

vertical stack: Mixing together of the records of several shots made in nearly the same location without correcting for offset differences. Used especially with surface sources in which the records from several successive weight drops, vibrations, pops, etc., are added together without making static or dynamic corrections to the components before adding (1).

Vibroseis: A seismic method in which a vibrator is used as an energy source to generate a wave train of controlled frequencies . . . Continental Oil Company trademark (1).

wavelength: The distance between successive similar points on two adjacent cycles of a monochromatic wave, measured perpendicular to the wavefront . . . (1).

well log: A record of one or more physical measurements as a function of depth in a borehole . . . (1).

well tie: Running a seismic line by a well so that seismic events may be correlated with subsurface information . . . (1).

window: 1. A portion of a seismic record free from certain disturbances, where certain important noise trains are absent. 2. The portion of a record chosen for designing operators such as those used for autocorrelation or frequency analysis. Also called gate . . . (1).

wipe-out zone: An acoustically transparent zone (i.e., without reflections) representative of surficial gassy sediments. It may also be caused by any other section without contrast (such as a mud filled channel) (4).

References

1. Sheriff, R. E., 1973, Encyclopedic dictionary of exploration geophysics: Society of Exploration Geophysicists, Tulsa.
2. Klein, M., 1976, personal communication.
3. American Geological Institute, 1960, Dictionary of geological terms: Garden City, N. Y., Doubleday & Co.

4. EG&G Environmental Equipment, 1976, Fundamentals of high resolution seismic profiling: Waltham, Mass., TR76–035.
5. Belderson, R. H. et al., 1972, Sonographs of the sea floor, a picture atlas: New York, Elsevier.
6. Baird, R. W., 1976, Prediction of potential drilling hazards by high resolution geophysical techniques: Continental Oil Co., Houston.
7. Bryant, W. R., 1984, personal communication.
8. Morgan, J. P., J. R. Morgan, and S. M. Gagliano, 1963, Mud lumps at the mouth of South Pass, Mississippi River; sedimentology, paleontology, structure, origin and relation to deltaic processes: Baton Rouge, Louisiana State Univ., Coastal Studies Inst., Tech. Rept. 10, 190 p.
9. Hedberg, H. D., 1974, Relation of methane generation to undercompacted shales, shale diapirs, and mud volcanoes. Am. Assoc. Petroleum Geologists Bull., v. 18.
10. King, L. H., and B. MacLean, 1970, Pockmarks on the Scotian Shelf: Geol. Soc. Am. Bull, v. 18.
11. Spencer, E. W., 1969, Introduction to the structure of the earth: New York, McGraw-Hill Book Co.

REFERENCES

Chapter 1

1. Fay, H. J. W., 1963, Submarine signal log: Portsmouth, R. I., Raytheon Company & Signal Div., 37 p.
2. Urick, R. J., 1967, Principles of underwater sound for engineers: 3rd ed., New York, McGraw-Hill, 342 p.
3. Del Grosso, V. A., 1974, New equation for the speed of sound in natural waters: J. Acoust. Soc. Am. v. 56, p. 1084.
4. Richard, J., 1907, L'Oceanographie: Paris, Vuibert & Nony, p. 23.
5. McClure, C. D., H. F. Nelson, and W. B. Huckabay, 1958, Marine sonoprobe system, new tool for geologic mapping: American Assoc. Petroleum Geol. Bull., v. 42, no. 4, p. 701–716.
6. Hersey, J. B., 1963, Continuous reflection profiling, *in* M. N. Hill, ed., The sea, vol. 3: New York, John Wiley & Sons, p. 47–71.
7. Fisherman's asdic mark II: London, Kelvin Hughes Co., Pub. M360, 1960.

Chapter 2

1. Dobrin, M. B., 1976, Introduction to geophysical prospecting: 3rd ed., New York, McGraw-Hill.
2. Grant, F. S. and G. F. West, 1965, Interpretation theory in applied geophysics: New York, McGraw-Hill.
3. Urick, R. J., 1967, Principles of underwater sound: 3rd ed., New York, McGraw-Hill, 342 p.
4. Bachman, R. T., 1983, Elastic anisotropy in marine sedimentary rocks: J. Geophys. Res., v. 88, no. B1, p. 539–545.
5. Del Grosso, V. A., 1974, New equation for the speed of sound in natural waters: J. Acoust. Soc., v. 56, p. 1084.

Chapter 3

1. Richards, A. F., ed., 1967, Marine geotechnique, preface: Urbana, Ill., Univ. Illinois Press.
2. Terzaghi, K., and R. B. Peck, 1948, Soil mechanics in engineering practice: New York, John Wiley & Sons, 566 p.
3. Lambe, T. W., and Whitman, 1969, Soil mechanics: New York, John Wiley & Sons, 553 p.
4. Scott, R. F., 1963, Principles of soil mechanics: Reading, Mass., Addison-Wesley, 550 p.
5. Wu, T. H., 1966, Soil mechanics: Boston, Allyn and Bacon, 431 p.
6. Bowles, F. A., W. R. Bryant, and C. Wallin, 1969, Microstructure of unconsolidated and consolidated marine sediments: Jour. Sedimentary Petrology, v. 39, no. 4, p. 1546–1551.
7. Milling, M. E., 1975, Geological appraisal of foundation conditions, northern North Sea: Oceanology International, no. 75.
8. Rock Color Chart, Geol. Soc. Am., Boulder, Colo.

9. Shepard, F. P., 1973. Submarine geology: 3rd ed., New York, Harper & Row.
10. Grim, R. E., 1962, Applied clay mineralogy: New York, McGraw-Hill, 422 p.
11. American Society for Testing and Materials, 1964, Procedures for Testing Soil: 4th ed., Philadelphia, ASTM, 535 p.
12. Denk, E., W. Dunlap, W. R. Bryant, L. Milleberger, and T. Whelan III, 1981, Pressurized core barrel for sampling gas charged sediments: preprint, OTC, paper # 4120, p. 43–47.
13. James, T. L., 1983, Gun-launched instrumented seabed penetrators: Initial field tests of the ISP–1 and ISP–2 Systems. Sandia Report: SAND 83–0095, Sandia National Laboratories, Albuquerque.

Chapter 4

1. Sverdrup, H. U., M. W. Johnson, and R. H. Fleming, 1942, The oceans, their physics, chemistry, and general biology: Englewood Cliffs, N.J., Prentice-Hall, 1060 p.
2. Del Grosso, V. A., 1974, New equation for the speed of sound in natural waters: J. Acoust. Soc., v. 56, p. 1084.
3. Collin Weeks, personal communication, 1983.
4. Multibeam bathymetric swatch survey systems, in Sea-Technology, June 1982, p. 28–31.
5. Dobrin, M. B., 1976, Introduction to geophysical prospecting: 3rd ed., New York, McGraw-Hill, 630 p.
6. Trabant, P. K., and B. J. Presley, 1978, Orca Basin anoxic depression on the continental slope, northwest Gulf of Mexico, in A. H. Bouma, G. T. Moore, and J. M. Coleman, eds., Framework, facies, and oil trapping characteristics of the upper continental margin: AAPG Studies in Geology no. 7, p. 303–311.

Chapter 5

1. Chesterman, W. D., P. R. Clywick, and A. H. Stride, 1958, An acoustic aid to sea bed survey: Acustica, v. 8, p. 285–290.
2. Belderson, R. H., N. H. Kenyon, A. H. Stride, and A. R. Stubbs, 1972, Sonographs of the sea floor, a picture atlas: Amsterdam, Elsevier, 185 p.
3. Urick, R. J., 1967, Principles of underwater sound for engineers: New York, McGraw-Hill, 342 p.
4. Flemming, B. W., 1974, Side scan sonar: a practical guide: Int'l Hydrographic Rev., v. 51, p. 1–27.
5. Flemming, B. W., M. Klein, and P. M. Denbigh, 1982, Side scan sonar techniques: Capetown, South Africa, W.G.A. Russell-Cargill.
6. Klein, M., B. D. Van Koevering, and F. C. Michelson, 1972, A new depressor and recovery method for undersea towed vehicles: 7th Annual Marine Tech. Soc. Conference preprints.
7. Franzen, A., 1981, HMS Kronan: The search for a great 17th century Swedish warship: Stockholm, Royal Institute of Technology.

Additional Bibliography on Side Scan Sonars
(Courtesy of Klein Associates, Salem, New Hampshire)

Bass, G. F., 1968, New tools for undersea archaeology: National Geographic, v. 134, no. 3.

Brooks, L. D., 1974, Ice scour on northern continental shelf of Alaska, *in* Reed and Slater, eds., The coast and shelf of the Beaufort Sea: Arlington, Va., Arctic Institute of North America.

Bryant, R. S., 1975, Side scan sonar for hydrography: Int'l Hydrographic Rev., v. 52.

Clay, C. S., J. Ess, and I. Weisman, 1964, Lateral echo sounding of the ocean bottom on the continental rise, J. Geophs. Res., v. 69, no. 18.

Coleman, J. M., H. H. Roberts, S. P. Murray, and M. Salama, 1981, Morphology and dynamic sedimentology of the Eastern Nile Delta Shelf: Mar. Geol.

Chapter 6

1. McClure, C. D., H. F. Nelson, and W. B. Huckabay, 1958, Marine sonoprobe system, new tool for geologic mapping: Am. Assoc. Petroleum Geol. Bull., v. 42, no. 4, p. 701–716.

2. Lowell, F. C., Jr. and W. L. Dalton, 1971, Development and test of a state-of-the-art subbottom profiler for offshore use: OTC preprint, no. 1340, v. 1, p. 143–158.

3. Blidberg, D. R. and D. W. Porta, 1976, An integrated, acoustic, seabed survey system for water depths to 2,000 feet: OTC preprint, no. 2655, p. 931–937.

4. Des Vallieres, T., H. Kuhn, R. LeMoal, and J. Duval, 1978, Test of various high resolution seismic devices in hard bottom areas: OTC preprint, no. 3221, p. 1455–1465.

5. Kozalos, J., personal communication, IDS (Nov. 1982).

6. Crawford, J. M., W. E. N. Doty, and M. R. Lee, 1960, Continuous signal seismograph: Geophysics, v. 25, p. 95–105.

7. Leenhardt, O., 1972, Le sondage sismique continu: Paris, Masson et Cie, 166 p.

8. Antoine, J., and P. K. Trabant, 1976, Geological features of shallow gas: Proc. Exploration and Engineering High Resolution Geophysics, Houston Geol. Soc., p. 21

9. Sieck, H. C., and G. W. Self, 1977, Analysis of high resolution seismic data, *in* Seismic stratigraphy—applications to hydrocarbon exploration: AAPG Memoir 26.

10. Prior, D. B., C. E. Adams, and J. M. Coleman, 1983, Characteristics of a deep-sea channel on the middle Mississippi fan as revealed by a high-resolution survey: Transactions, Gulf Coast Assoc. Geological Societies, v. 33, p. 383.

Chapter 7

1. McQuillin, R., M. Bacon, and W. Barclay, 1979, An introduction to seismic interpretation: Houston, Gulf Publishing Co.

Chapter 8

1. Hersey, J. B., 1963, Continuous reflection profiling, *in* M. N. Hill, ed., The sea, vol. 3: New York, John Wiley & Sons, p. 47–71.
2. Leenhardt, O., 1973, Le sondage sismique continu: Paris, Masson et Cie, p. 164.
3. Schor, G., 1963, Refraction and reflection techniques and procedures, *in* M. N. Hill, ed., The sea, vol. 3: New York, John Wiley & Sons, p. 20–39.

Chapter 9

1. Heitman, L. B., 1982, New technology enhances seismic search: Offshore.
2. Dobrin, M. B., 1976, Introduction to geophysical prospecting: 3rd ed., New York, McGraw-Hill, 630 p.
3. Seismic stratigraphy—applications to hydrocarbon exploration: Tulsa, AAPG Memoir 26, 1977.
4. SEG Technical standards committee, 1980, Digital tape standards: Tulsa, SEG, 65 p.
5. AAPG Studies in Geology #15, 1983, Seismic expression of structural styles, A. W. Bally, ed.: Tulsa, Am. Assoc. Petroleum Geologist.

Chapter 10

1. Packard, M. and R. Varian, 1954, Proton gyromagnetic ratio: Phys. Rev., v. 93, p. 941.
2. Bloom, A. L., 1960, Optical pumping: Sci. Am., v. 203, p. 72–80.
3. Bloom, A. L., 1962, Principles of operation of the rubidium vapor magnetometer: Appl. Opt., v. 1, p. 61–68.
4. Dobrin, M. B., 1976, Introduction to geophysical prospecting: New York, McGraw-Hill, p. 630.
5. Bullard, E. C., and R. G. Mason, 1963, The magnetic field over the oceans, *in* M. N. Hill, ed., The sea vol. 3: New York, John Wiley & Sons, chap. 10.
6. Breiner, S., 1973, Applications manual for portable magnetometers: Sunnyvale, Calif., Geometrics.

Chapter 11

1. A number of case studies are published yearly, in conjunction with the annual Offshore Technology Conference, which present a valuable source for HRG engineering case studies.

2. Milling, M. E., 1975, Geologial appraisal of foundation conditions, northern North Sea: Oceanology International, no. 75, p. 310–319.
3. Ardus, D. A., ed., 1980, Offshore site investigation, conference proceedings: London, Graham and Trotam, p. 290.
4. McQuillan, R., M. Bacon, and W. Barclay, 1979, An introduction to seismic interpretation: London, Graham and Trotam, p. 199.
5. Trabant, P. K., and W. R. Bryant, 1979, Submarine geomorphology and geology of the Mississippi River Delta front: Texas A & M University, Tech. Rpt. 79–1–T, 118 p.
6. Gagliano, S. M. et al., 1982, Sedimentary studies of prehistoric archaeologial sites: U.S. Dept. Interior, National Park Service.

Chapter 12

1. Gay, J. P., 1983, The role of navigation for the offshore petroleum industry: Sea Technology, March, p. 24–29.
2. Porta, D. W., 1980, An acoustic system for precision offshore navigation: Presentation at SEG Ann. MTG., Houston.
3. Stansell, T. A., Jr., 1983, Meeting the GPS challenge: Sea Technology, p. 48.
4. James, L. T., 1983, Gun-launched instrumented seabed penetrators: Initial field tests of the ISP-1 and ISP-2 systems: Sandia Report SAND 83-0095, Sandia National Laboratories, Albuquerque, p. 16.
5. Heubner, G., A. H. Bouma, and F. B. Chmelick, 1973, Sediment survey by high-resolve system: Offshore Technology Conference proceedings, paper no. 1845.

Additional Bibliography
(Courtesy of Klein Associates, Inc.)

Anonymous, 1982, Ice age plough marks a hazard to pipelines in northern seas: Offshore Engineer, May.
Bass, G. F., 1968, New tools for undersea archaeology: National Geographic, v. 134, no. 3.
Brooks, L. D., 1974, Ice scour on northern continental shelf of Alaska, *in* Reed and Slater, eds., The coast and shelf of the Beaufort Sea: Arlington, Va., Arctic Institute of North America.
Bryant, R. S., 1975, Side scan sonar for hydrography: Int'l Hydrographic Rev., v. 52.
Clay, C. S., J. Ess, and I. Weisman, 1964, Lateral echo sounding of the ocean bottom on the continental rise: J. Geophs. Res., v. 69, no. 18.
Coleman, J. M., H. H. Roberts, S. P. Murray, and M. Salama, 1981, Morphology and dynamic sedimentology of the Eastern Nile Delta Shelf: Mar. Geol., v. 42, p. 301–326.

Klein, M., C. Finklestein, 1976, Sonar serendipity in Loch Ness: Tech. Rev., December.

Klein, M. and J. Jolly, 1971, The use of side scan sonar to identify sea floor characteristics: Proc. Int'l. Symp. Engineering Properties of Sea Floor Soils and their Geophysical Identification, Seattle.

Klepsvick, J. O., and B. A. Fossum, 1980, Studies of icebergs, ice fronts and ice walls using sidescanning sonar: Annls. of Glaciology, v. 1.

Kozak, G., 1980, Side scanning sonar: A tool for the diving industry: Tenth Annual Int'l. Diving Symposium, New Orleans.

Leenhardt, O., 1974, Side scanning sonar—a theoretical study: Int. Hydrogr. Rev., v. 51, p. 61–80.

Lewis, K. B., 1982, Side scan survey of Taharoa Island and Terminal: New Zealand Oceanographic Institute Records, v. 4, no. 9.

Mittleman, J. R., and R. J. Malloy, 1971, Stereo side scan imagery: preprint, 7th Annual Offshore Tech. Conf.

Patterson, D. R., A. T. Shak, and M. T. Czerniak, 1982, Inspection of submerged arctic structures by side scan sonar: preprint, 14th Annual Offshore Tech. Conf.

Prior, D. B., W. J. Wiseman, and W. R. Bryant, 1981, Submarine chutes on the slopes of fjord deltas: Nature, v. 290.

Rusby, J. S. M., 1970, A long range side scan sonar for use in the deep sea: Int. Hydrographic Rev., v. 47, no. 2, p. 25–39.

Russell-Cargill, W. G. A., 1981, Developments in seabed mapping systems: Proc. Symp. on Hydrographic Survey in South Africa.

Spiess, F. N. and A. E. Maxwell, 1964, Search for the Thresher: Science, v. 145, p. 349–355.

Tucker, M. J., 1966, Sideways looking sonar for marine geology: Geo-Marine Tech., Oct.

Walsh, G. M., and G. J. Moss, 1970, A new approach to preliminary site surveillance: Navigation, v. 17, no. 2.

INDEX